The Earth Can Feed Us

The Earth Can Feed Us

BY

HUGO OSVALD

Introduction by

LORD BOYD ORR

Translated by

B. NESFIELD-COOKSON

THE SCIENTIFIC BOOK CLUB
121 CHARING CROSS ROAD
LONDON W.C.2

FIRST PUBLISHED IN GREAT BRITAIN 1966

SECOND IMPRESSION 1967

Translated from the Swedish
Jorden Kan Föda Oss

This second impression is published in
conjunction with George Allen & Unwin Ltd

PRINTED IN GREAT BRITAIN
in 11 *on* 12 *point Pilgrim type*
NOVELLO AND COMPANY LIMITED
BOROUGH GREEN, KENT

PREFACE

SINCE the Second World War my attention has to some extent been devoted to the problem of the world's food supply and its associated problems, a field of study very close to my activities as a teacher in matters of crop production, and to one who, in addition, has at several international meetings had the opportunity to discuss the problems of food supply for the poorer countries and their starving peoples. During a journey round the world between November 1951 and February 1952 I had the opportunity to see at close quarters the misery and the starvation, but also the measures being taken to increase food production and some examples of what can be done if science and technology are properly applied. As participant in the conference organized by the Weizmann Institute at Rehovoth, Israel, in 1960 on 'The Role of Science in the Advancement of New States', when much time was devoted to food supply problems, I had the opportunity of discussion with representatives of both rich and poor countries, thriving peoples and starving peoples.

On several occasions during recent times I have published articles on these problems in journals and newspapers. In these I have always, in the first place, tried to emphasize the possibilities of increasing food production by various measures but I have also refrained from giving terrifying descriptions of the situation in the over-populated countries, with their starving and diseased populations, or detailed accounts of the mistakes mankind has made in exploiting natural resources. For if we want to improve the situation, it is, in my opinion, necessary first of all to show what possibilities there are of increasing food production. Mankind must be given a gleam of hope, a chance to believe in a brighter future, if it is to make the necessary effort. Otherwise it will become discouraged and apathetic.

The increasing interest in many highly-developed countries in these extremely important problems, and their readiness to help the poor peoples to a higher standard of living is, in my opinion, a good reason for giving a comprehensive account of the whole complex of problems, of the possibilities there are of relieving starvation, distress and suffering – a fight which we must win if we are to avoid total catastrophe.

Since the Second World War a very rich literature has been published on these problems. Only a small part of that literature has been cited here.

In order to save time collecting, for the Swedish edition, the latest records from the different agencies of UN, at the beginning of 1962 I asked for assistance from my friend Sixten Heppling at the Swedish Agency for International Assistance, and this he kindly gave. Some time later, when he had to leave for Afghanistan, he referred me to his friend Mr Lars Eriksson, Secretary at the United Nations' Association for Sweden. Having read my manuscript, they both gave me guidance and valuable information. The publication of the Swedish edition was, however, delayed for more than a year, and a lot of new material, statistical records, results of scientific research and so on, had to be incorporated in the text. After the text had been revised, both Mr Heppling and Mr Eriksson read the manuscript again and gave me much information about the latest developments in international aid.

The Swedish edition was published in the autumn of 1963. Since then a lot has happened and it has therefore been necessary to revise the text once again. Sixten Heppling and Lars Eriksson have very kindly provided me with new material. In addition, Mr Mats Kihlberg at the Swedish Agency for International Assistance has provided valuable information about the Swedish activities in this field.

I appreciate very much the assistance these friends have given me and I take this opportunity to record my most sincere thanks.

HUGO OSVALD

CONTENTS

INTRODUCTION

LORD BOYD ORR

THE earth has never produced sufficient food for all its human inhabitants. With the development of agriculture the food supply has steadily increased, but the population has always increased faster, so people were always on the verge of a food shortage. Bad harvests which occur nearly every year in some part of the world have always been followed by famines. These occurred also in Europe up to the beginning of last century when the import of food from the sparsely populated countries of America and Australasia together with the emigration of millions of Europeans to these Continents enabled the increase in the food supply in Europe to surpass the increase in population. In other parts of the world, however, especially in India and China, famines continued up to the present century.

Today, owing to the great advance in agricultural science and in engineering, food is so easily produced in the western countries that there is ample food for everybody, with surpluses causing economic difficulties, and it is difficult for people in this and other western countries to realize that there is still a world food shortage.

In 1935 when the League of Nations promoted a World Food Policy based on human needs it was estimated that about half the people in the world lacked food sufficient for health. In 1946 when FAO submitted to all governments proposals for a World Food Policy it was estimated that, taking account of the rate of increase in population estimated at that time to be about 22 million a year, the world food supply would need to be doubled in the next twenty years to prevent a food shortage. Today, the world population is increasing at over 60 million a year. In the last few years the increase in the world food supply has barely kept pace with the increase in the number of mouths to be filled. Famines have been prevented in India by the generous supply of enormous quantities of wheat from America and minor assistance from other countries. China, owing to its great industrial developments and trade, was able to buy the surplus wheat of Canada and Australia and barley from Europe. Other countries have received help from the Rockefeller and Ford Foundations, organizations like Oxfam and the Freedom from Hunger Campaign. In spite of all this help to alleviate hunger, there are, today, more ill-nourished people in the world than there were before the last world war.

To provide sufficient food on a health standard for all the people in the world the present food supply would need to be nearly doubled, the biggest increase being in the more expensive foods such as meat, milk, fish, fruit and vegetables. By 2000 AD, i.e. only thirty-four years ahead, the present population of the world of 3,300 million is likely to reach 6,600 million, in which case the present supply would need to be increased four times if hunger in the world is to be abolished.

Some think that this is an impossible task, and the only solution of the difficulty is birth control to slow down and ultimately begin to reduce the number of people in the world. Birth control is urgently needed for other reasons besides food shortage. Doubling the present population and re-doubling it again by 2300 would raise grave social and political difficulties especially in overcrowded countries. The author of this important book has shown that it is physically possible to produce much more than four times the present food supply. Even if, as we hope, birth control succeeds in slowing down the population growth there is urgent need for producing sufficient food for all the people likely to be in the world in the next few years. When it is physically possible, two things are essential. The first is the enormous amount of money which is needed for the task of doubling the present food production. About £1,000 million would be needed in a co-operative plan by all nations as was proposed by FAO in 1946. That money could be obtained if the great powers would agree that all nations would cut their military budgets by, say, ten per cent as a first step to disarmament to save the world from a holocaust of destruction in a war with nuclear weapons. The half of that saving would give £1,000 million to begin the great task of freeing the world from hunger. The second difficulty is that the great powers which are so obsessed with their national interests and their fear and hatred of countries with different ideologies would need to realize that the only hope for civilization depends upon co-operation in removing the causes of war, the greatest of which at the present time is the revolt of the coloured races due to their poverty of which hunger is the first evil.

This book shows the dangers of world food shortage and makes a lucid statement of the possibilities of increasing food production to meet human needs. It should be read by all those interested in the kind of world our children will inherit, and especially by all politicians with the vision to imagine what state the world will be in fifty years hence. It is to be hoped that it will have a wide circulation.

BOYD ORR

ABBREVIATIONS

AID	Agency for International Development
DLF	Development Loan Fund
ECA	Economic Commission for Africa
EPTA	Expanded Programme of Technical Assistance
FAO	Food and Agricultural Organization
FOA	Foreign Operations Administration
IBRD	International Bank for Reconstruction and Development, or World Bank
ICA	International Co-operation Administration
IDA	International Development Association
IFC	International Finance Corporation
MSA	Mutual Security Agency
NIB	The Swedish Agency for International Assistance (Nämnden for internationellt bistand)
SF	Special Fund
SUNFED	Special United Nations Fund for Economic Development
UNESCO	United Nations Educational, Scientific and Cultural Organization
UNICEF	United Nations International Children's Emergency Fund
UNTAP	United Nations Technical Assistance Programme
WHO	World Health Organization

I

Can the Earth Feed Us?

THE question 'Can the Earth feed us?' may seem unwarranted to anyone who limits himself to a European horizon. Many European countries are self-supporting, at least in several of the most important foodstuffs. In fact, when yields are good, there is more than sufficient, and it is sometimes difficult to dispose of the surplus.

In the United States there is a similar situation, but even more pronounced. If American surpluses were to be thrown upon the world market, a catastrophic fall in prices would result, world trade in agricultural products would be completely disorganized and difficulties would arise for producers in practically every land. It is a known fact that, to avoid overproduction of grain, an attempt has been made in various parts of the United States to apply the principle of subsidizing farmers to leave a percentage of their arable land fallow. The same method has been used with other agricultural products. It was thought that in this way surplus production could be gradually eliminated. But this has not happened. On the contrary, the surplus is on the increase. According to an estimate made by FAO, there is likely to be during the five years 1962-66, a surplus of food to the value of about £4,500,000,000 in the United States alone.

It should be possible to use the surplus from several countries for international relief work (that is, apart from ordinary trade) with the object of improving the conditions in many underprivileged countries. This subject will be referred to later.

If then, we include the whole earth in our considerations, the question 'Can the Earth feed us?' is wholly apposite. In fact, we are here concerned with the most important international problem of our time.

The question can be answered in the briefest possible terms, as follows: the food supplies now available, or those produced annually, are not by any means sufficient, even if it were possible to distribute the apparently large surplus among the starving peoples of the earth. The sad truth is, that more than half the world's population of more than three milliard people has not enough to eat. Moreover, according to the investigation carried out by FAO, less than one fifth gets a sufficient quantity (60 gms a day) of high-grade proteins. Consequently the deficiency in calories is aggravated by the fact that the available food is often of inferior quality. Not only is it often lacking in essential amino-acids, but also in mineral salts and vitamins. It is, therefore, no exaggeration to maintain that about two-thirds of the earth's population are undernourished.

The question of the greatest interest for us all is, therefore: Is it possible to produce enough good quality food for everyone? Fortunately this question can be answered with an unqualified *yes*! But let me add at once that to this 'yes' is attached an inescapable condition; that mankind comes to its senses and utilizes the earth's resources rationally!

In the following pages I shall endeavour to show how, and by what methods the production of food could be increased so that everyone will be able to satisfy his hunger. However, the question of food production is closely related to that of the exploitation of a multitude of the earth's resources, and it will therefore be necessary to examine a few sources of supply other than those which can be directly utilized for the provision of food.

2

The Problem of Hunger through the Ages

MAN's first and greatest problem has always been that of satisfying his need for food. Clothes and dwellings, education and recreation, entertainments and enjoyments have always taken second place when bread has begun to run short. The struggle for life, as Darwin expressed it, has been and still is a struggle for *bread*, that is, for the means of subsistence. As Rosin (1953) said, people fight, not for the oxygen in the air, since its supply is limitless, but for such resources, essential for their food supply, which are, or are assumed to be available only in limited quantities – for example, fertile land, water, coal, oil.

For thousands of years, hunger – fear of hardship and privation – has been the most powerful driving force behind many of the events which have determined mankind's destiny. For instance, it has been assumed, probably rightly, that it was the deterioration of the climate during the transition from the Bronze Age to the Iron Age, and the consequent difficulties concerning food supply, which caused the migrations in the centuries about the beginning of our era. The French revolution of 1789 was certainly due to iniquitous social conditions, but it was the lack of food after the bad harvest of 1788 that precipitated events; we have a recent example in the revolution in Czechoslovakia, where the bad harvest in 1947 caused the food situation to deteriorate catastrophically, and thereby became one of the causes of the previous régime's downfall.

Throughout the history of mankind the shortage of food has determined the religious views of every nation. Behind the words of the prayer 'Give us this day our daily bread' lie the painful experiences of countless generations; and the exhortation 'In the sweat of thy face shalt thou eat bread' is an ancient truth which is still valid in many countries.

The reason for the overshadowing position of the food problem is that the population has always increased more quickly than man's ability to utilize and avail himself of the earth's riches. Within a limited area Nature has been generous now and then and made a gift of the abundance of the fruits of the earth – with a consequent sharp rise in population – but at the same time she has been parsimonious elsewhere. In earlier days when the system of transportation was little developed, such surpluses and shortages could be evened out only to a very limited extent. Seasonal changes expressed themselves, then, in strong fluctuations of the availability of food. Up to relatively recent times it was said of Sweden – as of many European countries – that there was a bad harvest every third year and a complete crop failure every tenth. A few lines from P. A. Säves's (1876) description of farming on Gotland in his interesting book *Tales from the Fields*, gives us an eloquent picture of conditions in the Swedish countryside in the eighteenth century: –

'In 1757 there was a fearful shortage of grain. Seed cost sixteen dalers a barrel and soon became unobtainable. In the following year, 1758, there was again an unprecedented scarcity on Gotland so that most people had to sell the greater part of their cattle, and the remainder collapsed with hunger and all the young lambs died. In 1760 it was again a bad year, the spring was so unusually warm that rye formed ears early, then on the 16th May there was a heavy snowfall followed by a violent flood which destroyed the whole crop. In 1763 there was a general shortage of fodder and it was difficult to obtain food, and also in 1769 when saw-sedge, lichens, spruce twigs and heather had to be collected for the cattle, and people had to take the ears of corn and dry them for food or cut the unripe corn. Then came the year of high prices (1772) when, moreover, Gotland was swamped by crowds of itinerant starving people from Oland who obtained food in exchange for copper pans, old-fashioned silver spoons and so on.'

Also in the literature of former times dealing with economy we can find all too many complaints regarding the insufficiency of the grain harvests. Grain was the most important food; even more so than it is today, for it was not as easy as it is now to supplement shortages from other countries. Nor was it possible to fall back on potatoes as we can now. The only ways out were to tighten one's belt and bake 'famine bread', of flour mixed with bark.

In his book *Ride Tonight* (1941), Wilhelm Moberg has given a vivid account of the conditions prevailing in the Swedish countryside during a year of scarcity:

'It was in the year of famine, it was a beggar's spring. The poor people came wandering from every direction, from known and unknown districts, both with and without the begging permit from the priest. – Never before had so many poor people crossed the threshold of Svedjegarden. Mother Sigga gave a morsel of bread to every one of the starving people as long as she could spare one. All the winter she had baked famine bread with chopped straw; the elm had to give its bark to the dough, and the hazel and heather their buds.'

With the improvement of transport and communication many of these difficulties were much eased for some European countries. On the other hand, conditions such as those just described existed in great parts of Eastern Europe well into the twentieth century. In many other parts of the world conditions are still much worse. For instance, poor crops in three consecutive years were the cause of the severe famine reigning in China a couple of years ago. Particularly in those parts of the world where the crops are dependent on the monsoon rains, it happens quite often that the rain fails and that many millions of people are threatened with death by starvation. An example of this will be given later on. Even if the geographical differences have been evened out by improved transport possibilities in many cases, those of an economic nature have remained, or have even been aggravated, with the result that in times of scarcity large population groups have had difficulty in satisfying their need of food.

This picture would be incomplete if I omitted to mention that there have been times of superfluity or, at least, of quite abundant supply – not for everyone, certainly, but for a large proportion of the world's population. After the American prairies became available for colonization hordes of people streamed to them mainly from Europe. The world production of foodstuffs grew rapidly and for about fifty years possibly half of mankind was relatively well off for food. But the consequent large increase in population, together with the rise in the standard of living in several countries, gradually reintroduced a shortage. In the meantime, the population increase in those countries which had not shared in the surplus from America caused a deterioration of the situation.

Towards the end of the nineteen-twenties and during the greater part of the nineteen-thirties there was again a period of apparent superfluity in Europe and America, followed by falling prices which created difficulties for agriculture in many countries. However, as everyone knows, it was not at all the case that the whole population of the earth could satisfy its need for food. In spite of the superfluity in certain countries at least one-third of the earth's population lived on the verge of starvation. But to ease the difficulties in the surplus areas measures were recommended and carried out to restrict production and stimulate consumption.

3

The Problem of the World's Food Supply

THUS, as we have seen, famine is nothing new. On the contrary, it is as old as mankind itself. But the problem is now far more serious than in former times, and this for two reasons.

In the first place, it is of quite a different magnitude because such multitudes are involved. Its magnitude can be indicated in a few words by quoting Paul H. Appleby, former Under-Secretary in the United States Department of Agriculture. In 1943 he described the situation in these tense and expressive words: –

'The starting point for any discussion of food is to be found in this basic fact: this long after the Industrial Revolution, two-thirds of the people of the world are normally engaged in producing food, and two-thirds of the people of the world do not have enough to eat.'

The situation has not improved since these words were spoken; if anything it has deteriorated. In 1960 there were 700,000,000 more people to feed than in 1940, but the production of food in 1960 was only a little greater than it was twenty years earlier. According to FAO,[1] food production per head is *now* about 13 per cent greater than it was before the Second World War, but, it is pointed out, this figure means little in a world in which production is so unevenly distributed amongst the population. Almost the entire increase is, in fact, in the highly developed countries – where, above all, an augmentation of stocks has been created – whereas in the underdeveloped countries the increase in production per head has risen by only 2 per cent.

[1] *The State of Food and Agriculture*, 1962.

In the second place, a large proportion of the starving, suffering peoples has come to realize that hunger and suffering, hardship and misery, do not need to be indissolubly bound up with human existence; that it is possible to raise the standard of living for everyone if efforts are made to make full use of resources, and if the rich peoples co-operate by placing capital for investments and technical knowledge at the disposal of the poor peoples. This realization of the possibilities of improvement has created the present state of political unrest in many parts of the world, and, as a result, existence has become more insecure for many of those nations who are, at present, well, or relatively well, off.

This attitude of the suffering nations was emphasized by President Truman in the Point Four Programme of the United States: –

'These people in recent years have been stirred by a growing awareness of the possibilities of human advancement. They are seeking a fuller life and are striving to realize their full capacities. They aspire toward a higher standard of living and better health and physical well being. Under present circumstances their poverty is not merely a handicap to themselves. By leaving them unable to fulfil their reasonable aspirations, their misery makes them fertile ground for any ideology which will hold out to them promise, however false, of means towards a better life.'

Men have long been aware of these dangers. Seneca the younger said in Roman times: – 'A hungry people will not endure reason, they will not listen to justice, and they will not bend to any prayers for mercy.' And in China there is an old proverb which says 'It is difficult to tell the difference between right and wrong when the stomach is empty.'

During the Second World War it was realized, particularly in the United States and in Great Britain, that a lasting peace could only be secured if it were made possible for the whole population of the earth to obtain food in sufficient quantity. To attain this goal it is not enough to be able to dispose of local, more or less chance surpluses – no matter how impressive they may appear to be in statistical records – but what is needed is a substantial increase in food production.

The situation in large parts of the world can be illustrated by an example from a country with a population of only 1.9 million people, Somaliland. After the war it was administered by Italy, but it had been decided that it should become independent in 1960, no matter whether the inhabitants – three-quarters of whom were nomadic – were ready for self-government or not.

In Somaliland, UNICEF has for some time been engaged in combating malaria. This work was begun in January 1956. At that time Somaliland had an infant mortality of 50 per cent. The children who survive are relatively immune to malaria, that is, they do not die of it. But between 50 and 80 per cent, depending upon the proximity of their dwelling-places to rivers and swamps, are infected.

In 1955 the *New York Times* wrote in a leading article (March 12th: –

'It is easy to think politically: we want all men to be free and trust them, if free, to survive. *But freedom is also freedom from disease and hunger*. Freedom from malaria in Somaliland can be had for about thirty-two cents a person. Freedom from hunger comes higher, yet, if oil and other resources are properly developed, may pay banking dividends as well as human dividends.

'Somaliland will not be free, in 1960 or at any other time, if disease among her people is not brought down to civilized minimums. *She will not be free until the specter of hunger is exorcized*. And these truths go for the whole undeveloped part of the world; here, in a relatively small area is the central problem of the future.'

4
Personal Impressions of Distress and Suffering

ONE can read book after book about the conditions in the over-populated countries, where, especially in the teeming cities, a large proportion of the inhabitants lives in utter misery, but it is nevertheless difficult to form a correct opinion of the gravity of the situation. This can only really be done if one has experienced the truly shocking distress at close quarters.

During a short visit to India in February 1952, I saw things that I shall never forget. I arrived in Calcutta by plane late in the evening. Already on the way into town from the airport I caught a glimpse of the meaning of over-population and of real distress. Masses of people were sleeping in the streets, even outside the entrance to the hotel where I was staying. Next day, driving about the city, I saw people collapse in the streets and remain where they fell, dying of starvation and privation. What struck me as being worst of all, perhaps, was that such occurences were so common that passers-by paid no attention at all.

In the ponds, on the outskirts of the city, the sacred zebu wallowed in the thick greasy mud, while both children and adults bathed close by, and women washed clothes in the filthy water.

Children sat in the streets and yards making 'cakes' of cow dung, which were then stuck up on walls and tree-trunks to dry and later be used as fuel.

If one has experienced these appalling conditions prevailing among such an ancient, highly gifted and civilized people as the Indians, the immensity of the problems involved and the gravity of their implications – even for us in the more privileged countries – are thrust upon one very forcibly. One cannot help but feel that it is our duty to render assistance – not, in the first place, to save ourselves but to assist our fellow-men in distress. I could not get to sleep for hours that first night, although I was exhausted by the long journey by air from Sydney.

A few days later I took the train to one of the places I had to visit. There were several Indians in my compartment. When they all got out at a station, one of them promised to get me a bottle of mineral water. While he was away a little girl came to me begging. She may have been five or six years old. I did not understand what she said, but I understood the appealing look in her great brown warm eyes and her gesture as she held out a little tin pot. But at that moment the man who had got me a bottle of mineral water came back. He pushed the little girl out of the way, and unfortunately, in the hurry of the moment, I had no coin handy. Just then the train started. The little girl gave me a long, beseeching, sorrowful look, that has remained fixed in my memory. For me, India's prayer for liberation from distress and suffering was concentrated in that look. That prayer must not die away unheard! It comes to us also from all other starving and suffering peoples.

5

Measures against Famine

THE devastation which European agriculture suffered during the Second World War, and the famine which followed it, must have made clear to most people that the whole earth was faced with serious problems of food supply. It was in order to try to solve these problems that the British and Americans decided to invite all the nations of the world to a conference at Quebec in October 1945 'to co-operate in abolishing hunger and raising the standard of living of land workers, the majority of whom in the under-developed countries live in abysmal poverty'. In this way the United Nations Food and Agriculture Organization (FAO) came into existence, and John Boyd Orr (now Lord Boyd Orr) became its first Director.

In October 1955, the tenth anniversary of the formation of FAO was celebrated at Quebec. 'Lord Boyd Orr renewed his faith in the organization's program with a prediction that by 1965 one would no longer see side by side in this small world piled up unsaleable food bringing depression in some countries and hunger due to food shortage in other countries.'[1]

This prophecy was no mere hope: it was a serious warning, amply justified. For only a few years after the end of the war, after the worst pressure had begun to recede and the prices of many necessities of life had begun to fall, people in many regions, including some of the war-ravaged countries of Europe, had begun to fear that the surplus crisis of the nineteen-thirties would be repeated.

Of course, we must not be led to believe that difficulties of the kind which were experienced in many countries during the nineteen-thirties will be permanently overcome. But equally we must not let such difficulties make us lose sight of the great problem – to increase the production of food. Local or chance

[1] *The New York Times*, October 1955.

surpluses must not lead to restrictions on production so long as a large proportion of the world's population is starving, but must be disposed of in such a way that they are used where they are needed.

The question of surpluses has been discussed in FAO on many occasions. It has been proposed, for instance, that a special organization should be created to take charge of the surpluses and to organize their distribution. At an FAO conference in Rome in November 1957, at which these problems were discussed, Professor Josué de Castro, a former Director-General of FAO, affirmed that the means at FAO's disposal were not nearly sufficient for the task. They were, de Castro said, 'only a drop of water in an ocean of misery'. It should be possible to effect a considerable improvement by food distribution. The aim of this would be not to allot supplies to all areas where, so to speak, constant under-nourishment prevailed but to any which had been subject to particularly severe famine, or for some other reason, were in need of temporary assistance with food supplies.[2]

One example may be given. In Pakistan, which had previously been self-supporting as regards food, the monsoon rains failed in 1952. Because of this a very severe shortage of wheat was caused, and the consequence was an acute famine. In 1952-1953 800,000 tons of wheat were imported to fill some of the gap, and for 1953-1954 even greater imports were needed. At the height of the crisis the situation was saved by the gift of a million tons of wheat from the stocks of the United States, and large contributions were also made by Canada and Australia under the Colombo Plan. For Pakistan, as for many other countries, the future depends on the speed with which irrigation systems can be extended.

The United States has intervened in similar catastrophes on other occasions by giving a part of their food stocks. But it would be a great advantage if all surpluses could be administered by an international organization.

If we can solve the great problem of increasing food production a great step will have been taken towards peaceful development in the world: if it is not solved, a third world war is a constant and imminent danger. Let us therefore examine the

[2] Concerning this problem see p. 118.

solution, the most important problem of our time. First of all we must turn our attention to the future evolution of world population.

6

Population Trends

SINCE pre-war days the population of the earth has increased by about 35 per cent – in the rich countries by about 21 per cent, in the poor by about 51 per cent – and at the beginning of the nineteen-sixties it was estimated at about three milliards: in 1964 it would be over 3.3 milliards. The annual increase in the population is at present estimated to be between 60,000,000 and 65,000,000, or about 170,000 a day. This means that the annual increase is about eighteen per thousand.[1] If it continues at the same rate the total population in ten years time should be over 3.5 milliards, and the daily increase, over 170,000. By the year 2000 the population of the earth can be estimated at 6 to 6.5 milliards.[2] It is this that has given rise to talk of a population 'explosion'. So if we are already faced with problems of nutrition, themselves already very difficult to solve, it is clear that the difficulties are going to increase rapidly.

To provide a really effective background to the talk of population 'explosion' statisticians usually calculate how large the population of the earth will be at a given moment in the future. For example, it has been calculated that by the year 2600 the density of the population will have increased to *one* person per square meter – that is to say, there will then be 'standing room only' on the earth.[3] Such calculations are of course futile, and their practical value as small as that of the calculations with which certain biologists have amused themselves in order to throw light on the reproductive capacity of various kinds of animals. For example, it has been calculated that if every cod's egg developed into a full grown cod, then in

[1] UN Population and Vital Statistics Report, 1962: Demographic Yearbook 1961. According to the Statistical Office, UN, the world population figure in mid-1963 was an estimated 3,218,000,000.

[2] Baade 1962.

[3] Sax, 1955.

only six years the whole Atlantic would be packed with cod from top to bottom.

In my opinion it is erroneous to talk of an 'explosion', but it is clear, of course, that population growth is now proceeding more quickly than formerly – for example, before the Second World War. The most important cause of this is that the campaign which the World Health Organization (WHO) has been conducting against such devastating diseases as, for example, malaria has been very successful in many areas.

7

The Need for Help

IN recent years it has become clear to more and more people in the Western World that a vigorous effort must be made to mitigate distress and suffering, and help the many millions who live in misery to attain a decent level of existence. But in what way shall this help be given? Many different proposals have been made in the last few years, and relief work has been done in different ways by a number of countries. The needs are of many kinds. Social care and medical care are needed; epidemics and devastating diseases must be fought; technical schooling, education and training, hospitals and schools, help with social reform, road-building, irrigation, etc. Many hold, and they are most certainly right, that it is necessary to stimulate industrial and commercial development. *But what is needed above all is food, more food, better food!*

When I visited India in 1952 and was able to inspect a great water-control project, the Damodar Valley Project, I discussed these problems with the director of the agricultural section of the enterprise, Dr Raman. This is the gist of what he said: –

'It is sometimes maintained that the first thing for us to do is to develop our industry. But even if we double our industry, indeed, if we multiplied it many times over, it is nevertheless a fact that by far the greater part of the population of India would be dependent on what the earth can give. And what the Indian people now need is, first and foremost, food.'

De Castro followed a similar train of thought, though he expressed it in a somewhat modified form: 'Under-developed countries which wish to develop industry must realize that the sources of energy, transport and food production must be developed together.'

It cannot be emphasized strongly enough and often enough that, of all the crying needs, the need for food must be satisfied first. Of the 54,000,000 deaths which occur every year, from 30,000,000 to 40,000,000 can be attributed to malnutrition.[1]

Those who have seen how things are in the over-populated countries, where multitudes are dying of starvation, are almost tempted to say that stones are being offered in place of bread if, for example, schools are being built for the starving peoples instead of assistance to produce food of good quality. Education is necessary, but first hunger must be satisfied! However, one of the most urgent tasks must be to teach the starving peoples how yields from the earth can be increased. Elementary education is, therefore, an essential element in the long-term campaign against hunger.

It is the same with the extraordinarily valuable work which is being done by WHO. It has been said – and not altogether without justification, even though it may sound cynical – that not much is gained by saving people from dying by disease only to give them, instead, a chance to die by starvation. It may seem paradoxical to work for the reduction of infant mortality and try to prolong the lives of great numbers of people, when at the same time so large a part of the population of the earth has not enough to eat and when tremendous efforts have to be made to increase the world's food production.

But as the Canadian Premier, Mr Pearson, said in 1955, 'The paradox is only apparent! A healthy world will also be a productive world. Disease, a high mortality and famine go hand in hand.' It is consequently a matter of the first importance so to organize relief work that the people who are saved from death through sickness also obtain food in sufficient quantity and quality.

Good food in sufficient quantity is not only an object to be sought in itself. It is also a factor in economic development. An interesting example of this can be found in Costa Rica. When the first Pan-American Highway was being built through this country, the workmen had a very poor capacity for work so long as their daily rations consisted of the usual diet of the Costa Ricans, which was deficient in proteins and contained a great deal of sugar. But after the diet had been changed so that

[1] de Castro, 1957.

it was nutritionally satisfactory, that is, contained enough protein, their efficiency was increased several times over. This result was so striking that the part of the road which runs through Costa Rica became known as 'the road that food built'.[2] Protein deficiency in food causes indolence, this prevents people, not only from working effectively, but also from availing themselves of measures to increase production which they would otherwise have been capable of taking. Kusum Nair (1961) has given many illuminating examples of this from India.

Lack of protein is also the cause of the serious disease called kwashiorkor, which affects children from the time they are weaned till about the age of seven, and which has long been a scourge in many tropical countries. The symptoms of this illness are, among other things, lack of appetite and stunted growth. In many parts of Africa it is reckoned that every person suffers from it during some part of childhood. Of the cases which have been brought to hospital the mortality till quite recently was 50 per cent. Serious difficulties of digestion are also among the symptoms. Many mothers have thought they could remove these by putting the children on a very scanty diet. But this only made the illness worse. The disease is worst in those regions where the children, after having been weaned, are put straight on to the diet of the adults, deficient in protein and too rich in starch.

The deficiency of protein and the consequent illness are particularly serious in Southern Nigeria, where manioc and yams are the principal foodstuffs. Here, however, there is now hope of preventing the illness by means of *amama*, a preparation manufactured in England. This is a powder which is put into the children's food and consists of peanut meal to which are added dried yeast, casein, sugar, five mineral substances and seven vitamins.[3]

The examples given show that there are great possibilities of using surpluses of proteinous foodstuffs to mitigate hardship and suffering, and increase working efficiency until the production of these things is possible in the distressed countries themselves. Here the northern countries too could give much help by supplying powdered milk and fish meal.

[2] Cépède, 1962.

[3] *Discovery*, Feb. 21, 1960.

8

Food Production or Birth Control?

IT has already been shown that (a) a successful fight against disease sharply accelerates the growth of population so that the food supply situation deteriorates still further, and that (b) an improvement of the food supply situation can also stimulate the growth of population, so that after a period of time the final result *can* be further deterioration. In such circumstances should not the first object be to retard the growth of the population, that is, work for birth control?

It must be obvious that an unchecked increase in population, with the alarming results that are revealed from time to time by the population statisticians, must lead to catastrophe and must therefore be prevented. Indeed, now and then, it is asserted that all help to distressed peoples is futile so long as it is not combined with, or preceded by birth control, as otherwise the relief only stimulates the population growth and aggravates the food shortage. When confronted with such statements, says Heppling, it is hard to avoid the reflection that it is striking how courageously some people bear other people's sufferings. As I understand the situation, the view the statisticians take is too one-sided. The bare fact that, at present, more than half mankind lives on the verge of starvation is surely proof enough that birth control alone cannot have the desired effect unless it is pushed so far as to reduce the world's population. And surely no one can believe that this is possible.

The object of our endeavours must consequently be to try, in various ways, to alter the relation between population and food resources, to try to reach a balance between population growth on the one hand and the increase in food production on the other, after the present shortage has been made good. To attain this result two things are possible: –

(a) to increase food production
(b) to increase production and to promote birth control.

There can be no question of advocating birth control by itself. Birth control has until recently played a subordinate part in the attempt to help the distressed peoples. The reasons for this are quite natural and obvious. Initiative in this sphere meets with too many obstacles in the form of religious taboos, traditions, deep-rooted customs, and so on; obstacles of a kind which do not affect relief measures that aim at increasing production. The initiative for controlling the birth rate must therefore come from within the over-populated countries themselves.

Such initiatives were taken, for example, in Japan and India at the close of the nineteen-forties. The resistance which has to be broken down in India can be illustrated by a little episode from my journey through the Damodar district with Dr Raman. We looked in at a small agricultural village where the clay huts clustered thickly together. Just as we were going through a narrow alley, a little girl of five or six came out of a doorway. She had the mark of her caste on her forehead to show that she was married.

'You see there,' said Dr Raman, 'what problems we have to struggle with. That girl is married already in accordance with tradition, and she will have children as soon as it is physiologically possible. That is why birth control measures are an important point in our five-year plan'.

There is every reason to assume that other countries will do the same as soon as possibilities of an improvement in the food situation are seen. But till then nothing is being done. Here again the indolence caused by lack of protein comes into the picture. 'When we are starving anyhow, it can't make any difference if there are a few children more or less,' was the attitude Kusum Nair often came across when she raised the question of family limitation in her talks with under-nourished people in Indian villages.

In most countries the high birth rate is probably due to the fact that a fairly large number of children is considered to be, and often is, an old age insurance for the parents, and also meets the need for a supply of cheap labour in agriculture. So long as infant mortality was abnormally high, the number of births was consequently very large. When the food situation has improved and infant mortality has diminished, the immediate result has been a large increase in population. Sooner or later, however, an improvement in the standard of living leads to a decrease in fertility. Even if, as has been said, no automatic connection between a rise in the standard of living and a decrease in fertility can be proved to exist, none the less it is a fact that such a connection does exist. Professor Baade in his book (1962) asks how long the present rapid increase in population will continue. His answer is that when prosperity and general education have become universal then the world population will either increase slowly or not at all – the exact opposite of what Malthus maintained. Our efforts should consequently be directed to attaining prosperity for all as quickly as possible.

Erland von Hofsten in 1962 expressed the same opinion. 'Rapid economic growth,' he writes, 'is the important and essential thing. There is nothing to show that economic growth would really go much faster even if we succeeded in reducing the birth-rate. Moreover, till cultural and economic development have started in earnest, it is idle to suppose that it is possible to reduce the birth-rate to an extent which will really have any significance.'

These views may be summarized as follows: measures of different kinds with the object of reducing the birth-rate are desirable, but the initiative must come from the people themselves. If expert medical help is asked for, then, of course, it should be given. But the rich countries should approach this question with the greatest caution.

There are, however, some who believe that more propaganda for birth control should be made by the wealthier nations. However desirable it may be that measures of this kind should be taken more quickly than at present, there is a danger that intensive propaganda would not only have no result, but would positively militate against its own ends. Among the coloured peoples propaganda for birth control is easily taken to be an attempt by the white man to preserve his power and authority. And is it really strange if, in these days, the peoples say, 'We will settle our own way of life. Don't interfere in our private affairs'? It is not least on this account that this problem must be approached with caution. But there are also other reasons which make this a 'sensitive' subject. Even in countries where birth control has been practised for a long time the correct term is avoided and supplanted by the diffuse and meaningless expression 'family planning'.

It may be added that in many quarters there are exaggerated ideas of the results which can be achieved by such propaganda, even apart from the type of obstructions just mentioned. Of the two milliard people who inhabit the overpopulated countries there are about 730 million in China. There, however, they are conscious of the problem. After a census in 1953 which showed that the population was larger than expected, intensive campaign for family limitation was started in 1957. But for some reason or other it came to a sudden end, perhaps because it was interpreted as an expression of communism's inability to produce sufficient food for the entire nation. Indeed, a couple of years later a high official denied that such a campaign had even occurred. 'It was only a question of protecting the mothers' health, nothing else, it was said.'[1] According to later statements from Hong Kong, however, efforts to promote birth control have been resumed, although with some reluctance. It is now directed at young marriages in particular. 'No man ought to be a father before the age of 26, and no woman a mother before the age of 23,' it is said. Consideration should be taken of the woman's ability to work. The reasons for the change in the population policy are not humanitarian but economic. Famine, the falling standard of living, expansion of the towns, and so on, raise the demand for a limitation of population-growth.

In China, then, there is no need for propaganda for birth control, and, even if there were, the Chinese leaders would hardly allow any interference from outside. Moreover, they are striving to extend their country's boundaries to include parts of more sparsely-populated neighbouring countries.

Several of the Arab countries also take a negative attitude towards birth control. In Japan family limitation has already been introduced, and in India, with a population of 450,000,000, there are, as already mentioned, provisions for family limitation in the national planning. The same is true of Pakistan and Ceylon.

[1] Wallmén, 1961.

In these countries, then, propaganda by the 'rich' countries is superfluous. Nor is it called for in Europe, North America, Australia and New Zealand. Consequently, any propaganda for family limitation could, today, only be directed towards a relatively small proportion of the earth's population – some 600-700 million. And a considerable part of this remainder lives in countries which are actually under-populated, for instance, large parts of Africa and South America.

To avoid misunderstanding it should nevertheless be emphasized that measures for reducing the birth-rate are urgent in many countries. The question is, really, how they should be applied in order to achieve results. As already mentioned, propaganda for family limitation is met, in many quarters and for many different reasons, with strong opposition. Dr Malcolm Tottie has also stressed that the problem of population can hardly be tackled in a supra-national way. One would have to follow a bilateral course as Sweden has done, for instance, in Ceylon, where – to use Dr Tottie's mode of expression – a Swedish group of experts is promoting family planning by going from door to door endeavouring to convince families of the importance, for both the individual and the community, of not having more children than they can support.

Dr Tottie points out that it is unreasonable to expect feelings of solidarity towards the community from people who live on the brink of starvation. What the field worker must stress again and again is that the health of both mothers and children will improve if the children do not follow one another too quickly, and that there is a greater chance of keeping three more or less well-fed children alive than eight who are permanently half-starving.

To this I would like to add that it will be easier to arouse interest in family limitation if, at the same time, it can be shown that measures have already been taken to improve the supply of food. As already emphasized, it is, above all, important to increase the quantity of protein as soon as possible.

Dr Tottie also rightly points out that, parallel with the measures for family limitation, steps must be taken to increase the production of food and to decrease the gigantic losses caused by a variety of parasites, for example, locusts and rats, 'which in some areas destroy half a year's growth of grain'.

Since the Swedish aid activity in Ceylon has been successful, other countries interested in obtaining help with population problems can be shown that 'the problem can be solved and that there are forces at work which are willing to give help – and soon'.

According to a communication from the present leader of the Swedish undertaking, Dr Arne Kinch, the project in Ceylon has been crowned with success. In one of the four districts where enlightenment and guidance have been given the birth rate has decreased by 30 per cent in four years. Dr Kinch says that this favourable result depends to a large extent on the fact that Ceylon has the best public health organization in Asia.

The problem of food production and population increase was discussed in some of the lectures given at the conference on 'The Role of Science in the Advancement of New States' at Rehovoth, Israel, in 1960.[2] The Principal of the Institute for Population Problems in Tokio, Professor Minoru Tachi,[3] gave an account of the population trends in Japan, and influencing factors. In 1933 the number of births was 33 per 1,000 and the net increase 16 per 1,000, resulting in a rapid increase in population. This reached a peak after the Second World War. But as a severe economic crisis then set in, illegal abortion became common. In 1948 abortion was legalized under certain stated conditions, and at the same time birth control advisory centres were established. The result of these measures was a marked decrease in the birth-rate. By 1959 the number of births had sunk to 17 per 1,000 and the net increase to 10 per 1,000. This is, perhaps, the best example we have so far of successful so-called family limitation. That it could be carried out so quickly depends, according to Tachi, partly on the general ability of the people to read, partly on the absence of religious taboos.

That literacy is a necessary premise for success in this kind of work has also been stressed by Alva Myrdal.[4]

Another speaker during the conference, Professor M. C. Shelesnyak,[5] Rehovoth pointed out that, *inter alia*, a great deal of

[2] See Gruben, Science and The New Nations, 1961.
[3] *Population Trends and Economic Development*: The Japanese Experience.
[4] 1961.
[5] *Biological Approaches to the Problem of Population.*

research is still needed into the interplay of biological factors and also into social and cultural problems. In his opinion, the new states should train the personnel needed to organize family planning.

In my opinion a very realistic approach was represented by Michael W. Perrin of the Wellcome Institute in London,[6] he argued on the premise that in future it will be necessary to have birth control so that a balance can be achieved between food supply and population, and also to counterbalance the increasingly successful fight against disease. However, since there are still many problems to be solved and many barriers to be overcome, it would be necessary for an interim period – which will probably be much longer than the enthusiastic advocates of birth control envisage – to direct efforts primarily towards an increase in the production of food, particularly foodstuffs rich in high-grade protein. Perrin pointed out that there are great possibilities in this respect, some of which are entirely neglected as yet.

The same opinion was held by Josué de Castro in a lecture which he had sent to the conference.[7] He stated that, owing to the recent progress in agricultural science, it is now possible to increase the production of food so that it outstrips the growth of population. Malthus' theory was hereby refuted. It is quite certain that, in the near future, it will be possible to achieve even more startling results than those which have been achieved in agriculture hitherto.

It will be possible, as James Bonner has expressed it, to produce food for a world population several times larger than at present. Owing to the great progress of agricultural science and technology it has become clear, according to de Castro, that the neo-Malthusians are talking nonsense when they say that birth control is inevitable if the world is to be freed from hunger. The population of the earth is far from being too large. We need not, therefore, 'be frightened by the fact that each day there appear, in the world, 80,000 new mouths to feed, since they are matched by 80,000 brains to think and create and 160,000 hands to work and produce'. What is essential is that the Earth's population makes use of its creative power to raise

[6] 'Food Production in Relation to Present and Future Population Figures'.
[7] 'Science and Technology in the Fight against Hunger'.

the standard of living by utilizing the progress science is making. Then Malthus' dilemma, the problem of producing sufficient food for an increasing population, will be definitely solved.

After the lectures on birth control and related problems an African said: 'Birth control may be justified in Japan, but it does not interest us.' This attitude is quite understandable, because many of Africa's under-developed countries are in fact not over-populated but under-populated. This is also true of other parts of the world, South America for instance. There could be no doubt that in the minds of those conference participants from the new states the thoughts of most, perhaps of all who were present, did not centre upon the possibilities of limiting the population, but upon those of increasing agricultural production, of combating devastating diseases, of improving instruction and education; thereby providing the basis for a more varied life and a richer spiritual culture.

However, there is a further aspect of family limitation, to which reference has already been made, which has a bearing on the problems of under- as well as over-populated countries; that is, the health of the mothers' and the welfare of the children. Once infant mortality drops from 80 per cent – at which high rate it actually stands in many countries – to a 'civilized minimum', it should be relatively easy to secure a hearing for the desirability of reducing the birth-rate. Many of the under-developed countries are sure to find soon that such a change lies in their own interest.

The doctrine held in many European countries that aid to the underprivileged countries must be combined with measures for family limitation can often prove a drawback to successful work. In the debate at Rehovoth concerning financial assistance an African expressed the opinion of many when he said: 'When 80 per cent of the population live on a minimal level in good years and starve in dry years, when infant mortality is 80 per cent and life-expectancy is half of what it is in Europe, when parasitic diseases weaken the whole nation, when human beings are forced to perform the work of animals, then there is absolutely no possibility of saving money in order to achieve a higher standard of living. Appeals for larger funds to help financially must be met with money, not with admonitions and preaching.'

It is the population statistician who presses the need for family limitation. The reason for this often seems to be that he does not realize the vast possibilities that exist of increasing the production of food.

9

How much does Food Production need to increase?

I HAVE already pointed out that the production of food is at present insufficient and that, as a result of the rapid growth in world population, it will decrease in proportion to needs unless it can be developed very considerably. We also have to reckon with the fact that future generations will assuredly demand not only an adequacy but a richer and more nourishing diet than that of earlier generations.

The question is: can we increase food production enough to satisfy such requirements? To answer this we must first find out exactly how great an increase is needed, and how present production is used.

Confronted with Appleby's statement, the task of satisfying the need for food seems hopeless. But if we look at the problem more closely it does not appear to be so overwhelming. According to a couple of American political economists, Pearson and Harper, in a book imbued with an ironic, sometimes almost cynical, negativism,[1] the average consumption of food per person per year throughout the world is about 250 kg. (600 lb.) of dry substance. According to these authorities 225 kilogrammes is needed to prevent starvation and 275 kilogrammes would be more than sufficient. Here, of course, dry substance is assumed to have average digestibility. The calorie value of 1 kg. (2.2 lb.) of dry substance can, however, be very variable, but since a considerable part of the food consists of cereal, the number of calories per kilogram could be put at 3,500-4,000. This means that the estimated quantities mentioned above more or less correspond to 2,250, 2,500 and 2,750 calories per day. These figures agree reasonably with a calculation made by FAO in 1946,

[1] Pearson and Harper, *The World's Hunger*, 1945.

according to which the average calorie-supply in the underdeveloped countries was lower than 2,000 calories, whereas the normal daily requirement is estimated to be 2,500.

The required amount of calories varies, however, according to the climate, body weight, occupation, and so on, and is therefore, as shown in Table I, different in various parts of the world. In a report published by FAO and UNESCO in 1962 the average requirement is estimated to be 2,400 calories per day. According to the same source the average consumption of calories was 2,070 per person.

Table I. REQUIREMENT AND CONSUMPTION OF FOOD IN DIFFERENT PARTS OF THE WORLD

	Group I	*Group II*	*World Average*
Calories, requirement per day	2,300	2,600	2,400
Calories, consumption per day	2,150	3,600	2,070
Total protein, gramme	58	90	68
Animal protein	9	44	20
Fat, gramme	34	106	56

Group I Far and Near East, Africa and Latin America, apart from the countries round the River Plate flood-plain.

Group II Europe, North America, Oceania and the countries round the River Plate flood-plain.

With such estimates as a point of departure attempts have been made to calculate how large that part of the earth's population is which is undernourished in terms of calories. Different authorities have arrived at different results. Thus, for instance, one authority has stated that, apart from China (not included in UN's statistics), some 10-15 per cent must be said to be undernourished. However, here, as Cépède has pointed out (1962), no consideration has been taken of unavoidable waste, that is, the difference between available and actually consumed food, which is probably in the region of 100 calories per day per person. Nor has account been taken of the circumstance that, in rich and poor countries alike, there exist smaller or larger groups which consume considerably more than the average quantity of food so that the remaining population has to be satisfied with a smaller consumption. Another authority points out that also in the United States a considerable part of the population can be said to be undernourished. Cépède concludes that probably about half of the earth's population must be regarded as being calorifically undernourished. This implies that quite a considerable improvement has taken place in comparison with the situation prevailing immediately after the Second World War. According to Parker (1962) the production of food per head has, in fact, increased by 14 per cent, but of this increase only an inconsiderable proportion has accrued to the under-developed countries. However, among these exist quite considerable differences. To mention but a few: In the Far East (apart from China) the food supply situation, in spite of a number of good harvests during recent years, has only been improved to the extent that the pre-war level has been reached. As the result of a series of natural catastrophes, drought, for instance, China has experienced a considerable decrease in production. In the Near East the increase per person is 10 per cent as compared with the pre-war level, but in Latin America this level has not yet been reached.

According to Mr B. R. Sen, Director-General of FAO, the population increase in the under-developed countries averages 2 per cent per year, whereas food production increases at the rate of 3 per cent – a positive difference which is all too little. Serious catastrophes, such as crop failure due to drought, and so on, can cause grave reverses. The year 1961/1962 was so bad, according to a FAO report,[2] that food production increased very little, if at all, and the available quantity of food per person was even less than in the previous year. However, the cause was principally poor weather conditions and may therefore be said to be transitory.

Tremendous efforts need to be made to increase production more than hitherto, especially in Latin America, Africa and the Middle East, where, according to Sen, the population in recent years has tended to increase more rapidly than the production of food.

To the malnutrition resulting from calorie deficiency must also be added that caused by a badly balanced diet – food poor in proteins, vitamins and mineral salts. Such deficiency gives

[2] *The State of Food and Agriculture*, 1962.

rise to what de Castro termed 'specific hunger', and is often the cause of diseases like scurvy, pellagra, beriberi, rickets and kwashiorkor (malignant malnutrition). Even a daily diet of 2,500 or more calories can be deficient if essential elements are lacking. Long before the symptoms of deficiency disease appear, such malnutrition, as already mentioned, is the cause of inefficiency at work, of apathy and indifference detrimental to production.

Consideration must also be given to the fact that all proteins are not of the same nutritive value. As is well known, they are composed of amino-acids of which some of the so-called essentials are lacking in most vegetable proteins, but are present in the animal proteins, that is, in meat, milk, eggs, fish, and so on. The required daily intake of protein is frequently stated to be 60 gms per person per day. If the total supply of protein in the food is taken into consideration then the required amount seems to be met, but, as is indicated in Table I, the supply of animal protein in the under-developed countries is, on an average, only 9 gms per day, and in one country it is only 4 gms.

If, then, consideration is taken to 'specific hunger', it is assuredly justifiable to claim that two-thirds of the world's population have not enough to eat. It is, as Cépède points out, a scandalous situation that, though it would be possible to nourish 35 milliard people and there are at present only 3½, nevertheless two out of every three people on this earth suffer from malnutrition.

By how much, then, does food production need to increase so that every human being can have enough good quality food? food?

As far as the calorie consumption is concerned the figures mentioned above show that an increase of 25 per cent would be sufficient. Over and above this, however, a considerable increase in the production of vegetables is necessary in order to produce enough animal foodstuffs. To produce 1 kg. of animal food 6-7 kg. of fodder is required (in both cases calculated as dry substance). The 'loss' is represented by the food required to maintain the animal. But if we remember that a completely uniform distribution of foodstuffs is impossible to achieve in practice, it would be reasonable to assume that the earth's vegetable production would have to be increased by 50 per cent to create a balance between need and supply. As the protein produced by cattle is replaced by fish products, the percentage figure mentioned above can be reduced. Subsequent to this the annual increase would need to keep step with population growth. If we lengthen our perspective to the year 2000, when the world population can be reckoned as twice what it is now, then it is clearly necessary to increase the production of foodstuffs threefold during the remaining decades of this century.

Even if the situation is not as frightening as it may appear at first sight, we are nevertheless confronted with a gigantic problem – even if we limit ourselves to the question of production and disregard the by no means easily solved problem of distribution. The risk of over-production is negligible. Harry Ferguson, the English tractor manufacturer, was certainly right when in 1947 he described the situation in the following manner: 'Gentlemen, I think that this answers any argument against an increase in production. If we do not do these things – increase production, reduce the cost of living, and provide employment for the people – we are going to have another war within twenty-five years, and nothing can stop it.'

Is it possible to increase the production of foodstuffs to the extent suggested above? This question has been discussed by many authorities since the war – Lord Boyd Orr, Fairfield Osborn, William Vogt, Pearson and Harper, Georg Borgström, Rosin and Eastman, Josué de Castro, Sixten Heppling, Fritz Baade, and many others. Even in those cases where the authors have not given any clear and unequivocal answer, their publications have nevertheless made a considerable contribution to clarifying the situation. Some of the authors mentioned take a pessimistic view, others look more optimistically at the possibilities of solving the problem. Rosin and Eastman, in their book *Road to Abundance*, suggest that we could solve all problems of food supply by utilizing the energy of the sun and staking very much on chemical research. On the other hand, Pearson and Harper show a very pessimistic attitude. The cynical negativism of these authors emerges clearly in the following statement: 'There is no shortage of plans and planners. There are those who would reduce disease. Others plan to end all wars, to prevent starvation and malnutrition, to eliminate fear and want, or to raise the standard of living. Still others plan for more children for future defence. Many of these plans are essentially in conflict with one another. *As in the past, wars, diseases and famine will be the major forces restraining the population.*' (Author's italics.)

Pearson and Harper were of the opinion that in the first place the universal problem of food supply should be solved by lowering the standard of living and reducing the population in densely inhabited Europe. When that suggestion was made in 1945 it must have seemed pure mockery to the peoples living in the war-devastated lands. How could such an idea have been realized anyhow?

In a UN publication of May, 1945, a number of experts scrutinized the possibilities of producing sufficient foodstuffs for the growing population, and they came to the conclusion that for a long time to come it will be possible to nourish a considerably larger population than at present, but that the actual development will probably go in the opposite direction. Their depressing prophecy is that ignorance and avarice, dissension and superstition, and tradition-bound prejudice will prevent man from starting the projects which he is able to accomplish, even though the alternative is destitution, starvation and suffering. Consequently the standard of living will sink.

Is this conclusion inevitable? I do not think so. A survey of the possibilities confirms what the experts have claimed: that from a technical point of view it is possible to effect such an increase that the population of the earth could have food in more than adequate quantity. But only on one condition: very considerable capital investment must be made with the purpose of (applying) modern agricultural technique everywhere. Lord Boyd Orr said in 1948: 'The carrying out of a world food plan would soon put an end to hunger, because, if science is applied quickly enough and with the same intensity as it was applied to destructive ends during the war, the loss of fertile soil can be stopped, production can be increased faster than population could possibly grow. The full application of science would double and redouble food production which would result in a higher standard of living and increased prosperity which is the only basis of world peace.'

Baade pointed out in 1962 that if the possibilities of increasing production are used to the full then food could be produced for more than 30 milliard people.

Such definite statements quite naturally provoke the question: how shall the increase in production be brought about?

Before answering this question, we should ask another: can we take better care of the produce the earth yields?

10

Can we improve the Management of the Earth's Produce?

Vegetable and animal foodstuffs

Foodstuffs are of two kinds: vegetable and animal. Plants are the predominant fodder used to produce the latter. In the transformation of plant into animal products only a fraction of the nutritive value (or calorie value) of the former is, as previously mentioned, recovered as animal food. The total supply of human foodstuffs is therefore strongly affected by the distribution between animal and vegetable foodstuffs. It is generally known that it is possible to nourish a greater number of people from the yield of a given acreage if they are exclusively fed on a vegetable diet than if they also consume animal food. We can illustrate this point with an example from the production of edible fat. To produce one ton of butter an area of about 12-12.5 acres is required, whereas the same quantity of vegetable fat can be produced from about 3.75 acres. As a result, of course, the price of butter is much higher than that of the vegetable fat. The supply of fat can then be considerably improved by a partial substitution of vegetable for animal fat. Correspondingly, animal protein can to a certain extent be replaced by vegetable protein, though it has to be borne in mind that in general the latter is not equal to the former in nutritive value. Only a few cultivated plants, for example the soya bean, have a protein content similar to that of the animal. Among other things, it is the quality of its protein which has made the soya bean such an extremely important vegetable for those peoples whose food consists largely of protein-deficient rice.

The reason for the inferiority of vegetable protein is that it does not contain certain essential amino-acids, lysine for example. By supplementing the vegetable proteins with these amino-acids – which can now be synthetically produced – they can be given the same nutritive value as the animal protein. Other foodstuffs, e.g. wheat and rice can, of course, be enriched with essential amino-acids in the same way. By treating protein-deficient rice in this way the standard of nutrition could be greatly improved in many districts, and serious deficiency diseases prevented. Such a programme could be quickly carried out and at very reasonable cost – only a few per cent of the production cost of the foodstuff.[1]

But even if, in certain countries, an increase in the consumption of vegetable foods and a corresponding decrease in animal foods were to effect an improved supply, this would play a relatively minor role in relation to the world food supply. Pearson and Harper record that only about 9 per cent of all foodstuffs is made up of animal foods. Large parts of the earth's population live – partly on account of religious convictions – almost exclusively on vegetable products. Among the numerous peoples of Asia and Africa animal products constitute only a small proportion of the food. We find the opposite extreme among the North American and Oceanian peoples.

In the previously mentioned FAO report, published in 1962, it is stated that with regard to its calorie value food is distributed among the various main groups in the following way: –

Table II

	**Group I*	**Group II*	*The World*
Vegetable food	88.5%	62.9%	78.8%
Animal food	6.7%	24.1%	13.3%
Fats and oils	4.8%	13.0%	7.9%

*cf. Table I, p. 44.

It is to be noted that the fats and oils comprise both vegetable and animal matter; predominantly vegetable in Group I, and probably more animal in Group II. In the Far East the percentage of animal food is no more than 5.7, and of fats and oils only 3.8. In North America the corresponding figures are 35.3 and 16.[2]

It must also be remembered that the attempts being made in many countries (e.g. Northern and Central Europe) to achieve

[1] Ericson, 1963.

[2] See *The World must eat.*

a higher standard of living for those in the lower income groups creates an increased demand for such animal foodstuffs as meat, milk and eggs. Because of this the improvement of the total supply of foodstuffs which could be obtained by such a measure as, for instance, substituting vegetable fat for a certain percentage of the butter, is more than outweighed. It can be taken more or less for granted that a rise in the standard of living will bring with it an increase in the proportion of animal products in the daily diet – even if certain changes of the proportion of the animal products may take place – and secondly, that the demand for an increased yield from the earth will consequently be more pressing.

Agricultural Products used to Produce Power

One factor which is often overlooked but which must be considered in this connection is the motorization of agriculture, even of the smallholdings. The greater part of the products which the earth yields are used for food, directly or indirectly. Only a minor portion of the total area is cultivated for other purposes, for instance, for the production of fibres for the manufacture of textiles (as cotton wool is replaced by synthetic fibres, so a considerable area can be made available for the production of foodstuffs), for other technical raw materials, for luxuries such as tobacco, and so on; but a very considerable portion is often overlooked – because it is not accounted for (in statistical records) in the quantity or value figures – that is, the production of 'animal power'. It is difficult to ascertain how large a part of the products are actually used to maintain draught animals, but quite clearly it is very considerable. Harry Ferguson, who on several occasions has drawn attention to this very point, has stated (no doubt with justification) that 99 per cent of the earth's cultivated land is worked by animal power. An intense motorization of the activities performed on the soil, and also of the various kinds of transport, would release enormous areas for the production of foodstuffs. The more we can decrease the area which is now being used for the production of power, the more we can increase that which is used for the production of foodstuffs. Motorization and mechanization both play a large and important rôle in many other respects in agricultural production, but that mentioned above is the largest and most important. According to Cooper, Barton and Brodell,[8] no less than 30 per cent of the increase in foodstuff supplies which occurred in the period 1920-1942 in the United States was obtained from those areas which had become available owing to the decrease in the number of draught animals. The remaining 70 per cent was due to the increased yield of both soil and animal and to the decrease of exports.

A similar development has taken place in other countries during the years since the Second World War. In Sweden, for instance, the stock of horses has fallen from 620,000 to 160,000. As a result a very considerable area has been liberated for the production of cereals and other vegetable products.

If all draught animals, horses, oxen, asses, etc., could be replaced by motor power it would mean, according to Baade, that the freed area (now used to produce animal fodder) could produce food for one milliard people.

Stock Losses and Waste

Of the foodstuffs which are annually produced – among others, approximately 980 million tons of cereal, 50 million tons of sugar, 275 million tons of potatoes – a great deal is wasted either on its way from the producer to the consumer or through defective storage. A. Ramsay Tainsh stated in 1962 that in India over 40 million tons of cereal are wasted every year by poor storage facilities, and that in other sub-tropical and tropical countries the losses are considerable. An investigation undertaken in twenty-seven countries had shown that the damage caused by insects, rodents and moulds to wheat, maize, rice, barley, oats, rye, beans and oil seeds accounted for losses of up to 54 per cent in some cases. Hitherto, Tainsh claimed, far too little has been done to rectify these bad conditions. He asserted that, instead, the poor countries have concentrated far too much on increasing the production of foodstuffs by such means as, for instance, irrigation systems. The solution to the problem of storage has been known for a long time: silos and air tight grain bins and, where required, treatment with insecticides. The cost of silos is relatively small, and it is therefore urgent to build them as soon as possible, thus enabling the many needy countries to make use of the grain from those with a surplus production.

[8] 1947.

'So much foodstuff is produced in the world today,' said Tainsh, 'that every human being could have sufficient to eat. The necessary work can be started as soon as governments, banks and businessmen realize that it will be cheaper in the long run to preserve and use the crops already harvested than to produce more foodstuffs to satisfy the noxious insects already well fed.'

The measures suggested by Tainsh correspond well with FAO's programme for the use of the surplus stocks in certain countries. But even if the supply of foodstuffs can be essentially improved by a considerable decrease in the losses of cereals, the need to increase the world's total production of food nevertheless remains.

It is also clear that the measures considered above do not solve all the problems connected with the storage of foodstuffs. The research now being done into storage problems in many countries may bring about a considerable and, in certain cases, fairly rapid improvement of the foodstuff situation. In this connection we may remind ourselves of the methods evolved during recent years to 'pack' milk in such a way that it remains fresh for several weeks. (See p. 96.)

In many countries oil seeds of different kinds are cultivated: soya beans, ground nuts, cotton-seed, and so forth. What remains after the oil has been extracted probably contains 20 million tons of valuable protein, only a small part of which is used for human consumption, the major part ending as animal fodder. A considerable part simply goes to waste. The oil is regarded as the essential product, the protein as a by-product. It would be better if the protein were used more intelligently, even at extra cost. To do so might present certain difficulties, but the quantity of protein involved is so large that the effort is worth while. Some experiments in this respect have been started in Central America and India and elsewhere. If they are successful, large quantities of 'waste' can become a valuable protein food.[4]

For several years now a few factories in England have extracted vegetable protein from a variety of raw materials such as ground-nuts, cotton-seed, grass, and so on, according to methods devised by Israel Chayen. The whole production, at

[4] Pirie, 1960.

present 1,750 tons per year, is sold to the foodstuffs industry.[5]

Another form of wasted foodstuffs are the green parts of plants – e.g. the leaves of the beet, which are either used as fodder or ploughed into the soil. They cannot be used as they are for human food, but the necessary apparatus already exists to separate the valuable proteins from the valueless plant fibres. Many plants could undoubtedly be cultivated for the double purpose of producing protein as well as another valuable substance, for example, oil, fibre or sugar (Pirie 1960).

Very large quantities of food-substance from animal products also go to waste, in the form of slaughterhouse refuse for instance. By a method developed at the microbiological department of the Royal School of Agriculture at Uppsala, this refuse can be silaged together with cheap carbohydrates and malt as added ingredient. The silage thus obtained can be used, in its dried state, as food. Fish refuse can be treated in the same way to give a valuable additional supply of protein, particularly along the coasts of Asia and Africa.

[5] Batchelor, 1963.

II

Ways of increasing Food Production

THE total product of the earth can be increased in many different ways. For instance, the cultivated area can be enlarged; or the yield per area can be increased, by improving the conditions of cultivation, by water supply, providing plant nutrients, by better methods of cultivation, and so on, or by improving the cultivated plants and efficiently controlling diseases and pests. In many places cultivated land is subject to destruction, yet measures can put a stop to it. In many countries the arable land is so split up that effective cultivation is prevented. In such cases redistribution of cultivated land is necessary.

A universal programme for increasing the production of foodstuffs would have to embrace, among other things, the following: –

(1) Bringing hitherto uncultivated regions of the earth into use.
(2) Taking steps to prevent erosion.
(3) Redistributing cultivated land in many countries.
(4) Controlling the supply of water either through irrigation or through drainage (in many cases both are needed).
(5) Supplying the soil with calcium and plant nutrients (artificial fertilizers).
(6) Applying modern cultivation techniques.
(7) Mechanizing agricultural work; among other things, reducing the costs of production.
(8) Plant breeding to improve the cultivated plants in all countries in yield, quality, resistance to disease, etc.
(9) Controlling plant diseases, pests, weeds, etc.

Several of the measures just mentioned can be furthered by

economic means, by co-operation, by better marketing organization, by endeavours to bring about a fairly stable price level, and so on. Here it should be emphasized that an essential condition for the realization of the programme is that those engaged in agriculture should get the necessary education and training and that consultants and instructors should be employed in all the under-developed countries. The battle against starvation must be conducted side by side with the battle against illiteracy and ignorance.

Furthermore, it should be remembered that provision must be made for those who gradually become redundant as agriculture is rationalized and machines replace animals. Work will have to be found for them in other occupations, so that industrial investment will be needed, primarily for manufacturing the means of agriculture production, for instance, artificial fertilizers, and for the processing of the earth's produce.

Can the cultivated area be increased?

The possibility of increasing the cultivated area of the earth was realized in the nineteenth century when a considerable proportion of Europe's surplus population went to find its livelihood on the American prairies. Is a similar solution feasible today?

There is, of course, no continent today that is entirely undeveloped. But are there possibilities of increasing the cultivated area on those continents which are already more or less cultivated? This question has met with a great variety of answers by different investigators. It appears that, until very recently, such possibilities have generally been regarded as relatively insignificant. However, it seems that such a judgment is very largely influenced by Pearson and Harper.

Let us look, then, at the problem quite briefly. The cultivated area of the entire earth amounts to about 3.25 milliard acres or about 10 per cent of the earth's land surface. According to Pearson and Harper: 'There is no scarcity of land with favourable topography, or adequate sunlight and carbon dioxide, or fertile soil, or favourable temperature, or reliable rainfall. There is however, a serious shortage of land with the proper combination of these various factors that are so necessary for food production.' However, other authors have arrived at more encouraging results. For instance, Baker (1923) estimated that the area suitable for wheat alone, amounted to 3.75 milliard acres, and if the areas on which other cultivated plants, but not wheat, could be raised are added to this a considerably higher figure is arrived at.

Estimates made during recent years indicate that nearly half of the earth's land area could be cultivated. In these estimates are included the tropical belt, savannah and monsoon regions in Asia, Africa, America and Australia, certain regions in the temperate zones, but not the deserts.[1]

In the countries near the equator – that is, within the tropical rain forest belt – the conditions for agriculture, or, rather, for certain branches of plant cultivation, are relatively unfavourable. On account of the abundant rainfall and the high temperature, the top layer of the soil has become impoverished – through the so-called laterite weathering – and if the ground is left open in these districts, after the forests have been felled, it is subjected to severe erosion. The shortage of plant nutrient can, however, be easily remedied by fertilizers, and if arboreal plants are cultivated, such as the rubber-tree, cacao-tree, oil-palm and the *Citrus*-species, then erosion can be avoided. In the Congo the so-called corridor system has been developed in some quarters. This means that quite narrow strips of land are cleared and cultivated, between which the primeval forest is left standing. After a few years of cultivation the strip is allowed to revert and a new corridor is opened up. In the future the major part of the tropical forest regions will no doubt be best made use of by rational forestry. But even if only 20 per cent of the unused tropical rain forest regions were to be used for the production of foodstuffs, 1,000 million acres would be gained.[2]

There are also enormous possibilities for rational cultivation in the savannah and monsoon regions of the earth. These are characterized by a longer or shorter period of rain for a part of the year, and by an intense drought for the rest of the year. As an example, the broad belt in Africa south of the Sahara, from the Atlantic coast to Abyssinia, might be mentioned. Large parts of these regions are used as pasture-land but could be used far more usefully for the production of maize, ground-nuts, oil-

[1] See, for instance, Nilsson-Leissner, 1955.
[2] Cf. Nilsson-Leissner, 1955.

plants, and so on. But the condition for such a development is large capital investment for the regulation of water supplies. In the Soviet Union, it is said that plants are being constructed to irrigate 20 million acres in the steppe regions, and as a result of the fantastic Ob-Yenisei project some 62.5 million acres of arable land will be gained. Two great Siberian rivers, the Ob and the Yenisei, will be dammed up and a considerable part of their waters will be led southwards instead of northwards; an inland sea will be created in Central Siberia, and it will be possible, among other things, to irrigate the Turkmenistan steppe. This steppe, says Baade, is called 'starvation steppe' because nothing of nutritive value can be grown on it owing to the lack of water.

Large parts of the steppe and monsoon regions in Asia are already being used for the cultivation of different crops, but because of drought these areas lie fallow for more than half the year. They could be used far more effectively if irrigation were provided. The limiting factor here is not lack of knowledge but lack of money.

Arid and semi-arid regions can be made use of if irrigation is possible, a question to which we shall return later.

In the temperate zones, too, there are considerable cultivable reserves in the Soviet Union as well as in North and South America. West and Central Siberia comprise an extremely interesting zone; there the permanent frost is retreating and cultivation consequently advancing to the extent of half a mile or even a mile a year.[3]

As already mentioned, however, in many cases, large and well planned projects are called for to deal with drainage, irriga-

[3] This development in Siberia is an example of a very interesting geological phenomenon. During the long period of time in which the whole of northern Europe and parts of Siberia were covered with ice, the frost penetrated to great depths in large parts of Siberia. After the ice had receded the frost began to thaw, and soon the soil was covered with vegetation. The lower, still frozen levels were isolated as a consequence, and thawing progressed, therefore, very slowly – particularly as the winters in these regions are extremely cold. As long as there was a frozen layer at great depth below the surface, the upper layers could not become sufficiently warm during the summer to make cultivation possible. It has taken thousands of years to thaw the frozen depths, and what is now happening is that a strip is being regularly freed from the coldness beneath; the upper layer can be warmed sufficiently during the summer for the soil to be cultivated: *Discovery*, 6:2. 1945.

tion, and other such measures. Mention may be made here of the vast irrigation system which has been carried out below the Grand Coulee Dam in the state of Washington. There the waters of the Columbia River are used to transform vast areas of desert-like steppe into fertile soil. Another large water-regulation undertaking is the Indian Damodar Valley Project, by means of which it will be possible to irrigate an area of about 1,500,000 acres.

A very large project which has received much publicity for a number of reasons is the Egyptian plan to regulate the Nile. This involves, first and foremost, the building of the enormous Aswan dam, which will put a large area under water. By this means not only will the yields of the whole Nile Valley be increased and made reliable, but it will be possible to extend cultivation over large, arid areas hitherto unusable.

Early in 1960 it was mentioned in the press that a definite agreement had been reached on the final shaping of one of the largest irrigation projects in the world: that is, regulating and using the enormous volume of water of the river Indus. This solved the twelve-year old dispute between India and Pakistan: how the water resources should be distributed between the two countries. The project is estimated to cost an enormous sum, but it will directly affect 40 million people.

Mention may also be made here of the projects which the Chinese People's Republic intend to undertake in order to tame their great rivers from their sources to their mouths, above all the Yang-tsi-Kiang, Huang-ho and Si-Kiang. The Huang-ho, for instance, is to be dammed up in forty-six steps, whereby nearly as much electric energy will be generated as is – from water power – obtained in the whole of Europe. But water power is only a by-product, the main purpose of the project being to prevent flooding on the one hand and to secure water for irrigation on the other.

Further examples of such undertakings are the Gosina Project in the Sudan, and the regulation of rivers in Nigeria, Volta and Indo-China.

In the future it will also be possible to transform enormous desert areas into fertile soil – a problem we shall consider later (see page 73) – and thereby the acreage of cultivated land can be increased quite considerably.

By the regulation of rivers and the cultivation of steppe areas, which are now partly used as pastures, of areas which in FAO's statistical records are characterized as not utilized but fitted for production, of some part of the tropical rain forests and jungles and of some desert areas, it might be possible to increase the cultivated area to nearly 10 milliard acres, that is about three times as large an area as the present one. However, although it is possible to enlarge the area of cultivated land quite considerably, it is rather probable that the acreage of cultivated land will increase at a slower rate than does the world population. The fact is that an increase of the cultivated land of the earth by two per cent per year would mean that every year 67.5 million acres should be made use of. Nor is it necessary to avail oneself of all the possibilities to cultivate new land, since an increased production can be obtained at a lower cost by measures to increase the yield per unit area.

Measures Against Soil Erosion

Soil erosion, involving its destruction by water and wind, constitutes in many places a serious threat to food supplies. It is just as important to check this process as to cultivate new soil. Erosion is usually the result of faulty methods of cultivation in the past or present, or of over-grazing. Lord Boyd Orr said in a paper published in 1948: 'It takes nature from 300 to 1,000 years to build up one inch of fertile soil. Man by his wanton mis-use can destroy eight inches in one or two generations.' Sometimes much more. In former times the destruction of the soil has impoverished fertile lands and has been the cause of the downfall of civilizations.[4] Today it is a more serious problem than ever. In America alone three milliard tons of fertile soil were lost annually before an intensive campaign was begun to prevent these losses. According to Yudkin, some 300 million acres must have been destroyed in 100 years in the United States, and a considerably larger area damaged to some degree.

Perhaps China offers the most striking example of the misuse of the earth. There the forests have been devastated and agriculture pursued in a way which has caused very serious erosion. Yudkin states that 25 per cent of the cultivated or cultivable land must have been destroyed. Furthermore, it is reported that

[4] Cf. Boyko, 1947.

the Yellow River carries away 25 million tons of good soil annually. The country is now confronted with a problem which is very difficult to solve as the annual increase in population is 20 to 25 million.[5]

In all endeavours to conserve the soil it is, above all, important to devise an agricultural procedure which prevents water and wind from causing deterioration. Included here is also the problem of preserving the soil's humus. This problem is equally serious in many European countries.

Erosion wreaks havoc in both arid and damp zones, whether the agricultural method is developed or not. In Sweden, for instance, some areas with sandy soils have been subjected to severe wind erosion, which has not only spoiled large areas of fertile soil but also spread the potato nematode over large districts from small infected areas.

Attempts have been made, occasionally with success, to prevent erosion by wind by planting hedges as windbreaks; other cases have called for protective belts of trees. As a general rule, the cultivation of the soil and crop rotation should be so planned that the soil is covered with growth to the greatest possible extent during that time of the year when the risk of wind erosion is greatest.

I gained a very strong impression of the rapidity of soil destruction during a visit in 1951 to a pasture district in New Zealand. The forest had been felled and pastureland prepared in a very hilly landscape where the earth's upper layer was a sandy soil on loose sandstone. After the forest had been replaced by grassland the run-off had quite naturally increased and then erosion had set in. It is true that the luxuriant and dense grassland cannot be cut to pieces by the rain-water, but erosion works from beneath; the soil becomes saturated and then, quite suddenly, a part of the grass-grown slope slides down, leaving a large and gaping wound. Thus the erosion gains ground, biting deeper and deeper into the slope, and in a short time a ravine is formed. The creek which results from such erosion is not only a threat to the pasture-lands but also to roads, buildings, and so on. In parts of this district the running water had carved ravines in the loose sandstone ten to fifteen metres deep. We came across some steel wire hanging across one of these ravines, and

[5] *Demographic Year Book*, 1961.

my fellow-traveller told me that five years earlier it had been a fence between two pastures. Attempts are being made to stop such erosion. For instance, palisades are set up in the ravines to arrest the flow of sand, and by gradually building up terraces to arrest the spear-head of the erosion. However, many people are of the opinion that in such cases there is only one cure: to replant the forests over the whole area. A good survey of the different forms of soil erosion in New Zealand has been given by Grange and Gibbs.

In the relatively dry districts within the monsoon regions it occasionally happens that the monsoon rains are unusually heavy. In such circumstances even otherwise insignificant watercourses can cause considerable damage to cultivated fields. I have seen examples of this in India. Fertile plains can be severely cut about. Quite often, however, the damage is not so bad but that it can be remedied by levelling. Large quantities of top soil are lost nevertheless. One way to prevent such damage is to control the flow of water by building a number of small regulation dams. The water is collected in these dams during the rainy season and led out over the neighbouring fields during the dry season. The latter are provided with low walls which closely follow the contours of the hillside; through these the water seeps, trickling from one side of a field to the other. In this way the water is utilized intensively, but is prevented from causing damage. In some cases, contour ploughing, strip-cropping and so on, can prove fully adequate in preventing the top soil from being washed violently downhill.

Many measures against soil erosion are closely connected with the control of water as I have just made clear (see p. 58). A useful opportunity to discuss these problems occurred during the Rehovoth Conference, when Professor Lowdermilk of California gave an interesting survey of the steps that must be taken in the new states, as well as in some of the more advanced countries, to achieve a rational management of soil and water. The concluding words of Professor Lowdermilk's lecture gave vivid expression to the importance he rightly attributes to the care of the soil. He said that when he had seen how mankind had destroyed the productivity of vast areas of the earth, reducing nations, great and small, to poverty and decline, he had often wondered if Moses, had he been able to foresee the wastage and neglect of the earth, might not have been inspired to deliver an eleventh commandment to control man's relation to the Earth.

In a radio address from Jerusalem in 1939, he suggested the words of such a commandment and dedicated it to the Jewish agricultural colonists, whose good stewardship was his inspiration: 'Thou shalt inherit the earth as a faithful steward, conserving its resources and productivity from generation to generation. Thou shalt safeguard thy fields from soil erosion, thy living waters from drying up and thy forests from desolation, and thou shalt protect the hills from over-grazing by thy herds, that thy descendants may have abundance forever. If any shall fail in this good stewardship of the land, thy fruitful fields shall become sterile, stony ground or wasting gullies, and thy descendants shall decrease and live in poverty or perish from the face of the earth.'

Redistribution of Cultivated Land

In many countries, among others India, the cultivated land is divided into such small plots that modern agricultural techniques cannot be used. During my visit to India I had occasion to see maps showing the division of the land in some villages west of Calcutta. It was like seeing a map of a Swedish rural village in the eighteenth century, except that the land division was even more drastic. The majority of the peasants in the early nineteen-fifties had extremely small parcels of land at their disposal. Units of up to five acres accounted for 88 per cent of the farms in the province of Uttar-Pradesh, while in the Punjab the corresponding figure was 64 per cent, and in Bengal 71 per cent. In Madras 74 per cent of the farming units were no larger than 2.5 acres.

In these areas a redistribution of the cultivated land and a merging of fragmentary units is essential to the success of other measures to promote production. In a country like India this is in itself already a gigantic undertaking requiring a large capital investment, but work to this end has already been going on in a number of districts for ten years or more.

More deplorable than this minute division of cultivable land is the fact that in many of the poorer countries it is owned by a few great landowners who live in luxury and affluence while the land is cultivated by poor tenants who not only have to pay extortionate rents – up to 80 per cent of the annual yield – but are also virtually without legal rights. Consequently, the practising farmer has no interest in taking expensive measures to increase production.

Kusum Nair has given us a clear description of the state of things prevailing in the farming structure of India. In the Tanjore district, of 575,000 landowners only 255 own 125 or more acres. 504,620 owners have 3.75 acres or less, and 401,741 1.25 acres or less. But none of these landowners devotes himself to crop production, which is the work of landless labourers or tenants. By law the latter must hand over 40 per cent of the harvest to the owners, though in practice it is often as much as 60 per cent. It is not surprising, therefore, that the farmers take no interest in increasing production, nor have the means to do so. One peasant, with whom Nair conversed, rented 1.25 acre and had to provide for a family of six. He normally produced twenty sacks of rice annually, of which he kept twelve. 'I don't need any more, that is more than enough,' he said. As long as the farmers take this attitude, propaganda aimed at an increase in production will be of little avail. Social conditions must be changed to give farmers an interest.

Progress is impeded also by the fact that the poorer peasants have little chance of securing credit, and, as a result, are in the hands of private money-lenders – usually the landowners themselves – who claim excessively high interest rates on loans. An investigation made by the Reserve Bank of India in 1954 showed that only 6.3 per cent of the credit required by the farmers was provided by the state and by the co-operative societies, the remainder being provided by other lenders at unreasonable interest rates. An investigation of agricultural credits, sponsored by Sweden, has been made by an FAO group of experts, divided into two teams, working in Asia and Africa respectively. Very valuable material has been provided concerning ways to strengthen the provision of agricultural credits in underdeveloped countries, and for the systematic planning of land tenure. In the comprehensive report delivered by the Swedish member of the group, reference was made to the special conditions existing in Latin-American countries. Measures based on this report for assistance within the programme of the FAO are now being discussed. To begin with, assistance will probably be concentrated to two or three countries in Africa or Asia.

In the advanced countries the problems connected with the division of land and with land-ownership have scarcely been considered in discussing assistance to the poor countries, though they have been comprehensively discussed by Peter Ussing Olsen. A few examples may be mentioned here to clarify the situation. In Latin America 1.5 per cent of all estates have more than 15,000 acres each, and this small percentage comprises half the total area of land under cultivation. The El Carmen estate in Bolivia consists of properties totalling five million acres.

In Brazil the distribution of land appears as follows:

Table III

Size of farm in acres	*Per cent of total number of farms*	*Per cent of total area of farmland*
Under 25	34.5	1.3
25-250	51.0	15.3
250-2,500	13.0	32.5
2,500-25,000	1.5	31.5
Over 25,000	0.1	19.4

Though conditions vary considerably from one country to another, it is nevertheless generally true that the farming population of the poorer countries lives at a very unsatisfactory level, not only in an economic sense, but socially, too. Western aid to the poorer peoples should therefore not be directed exclusively towards technical assistance of different kinds, but should also aim at supporting those measures aiming to effect land reforms of such a nature that the farmers will be interested in, increasing the production of foodstuffs and will have the means to carry them through. A land reform policy in conjunction with the redistribution of cultivable land is, in many under-developed countries, quite simply an absolute condition. Unless this is met, the peoples of these countries will not be able to use the possibilities of increasing the production of foodstuffs which will be at their disposal as production is technically improved.

The Yield per Area Unit

In the preceding sections we have dwelt on the possibilities of increasing the cultivated area, of protecting the land already under cultivation against destruction, of reducing storage losses and of having the land better distributed. Before proceeding to a consideration of the possibilities of increasing the yield per area unit, we may first look at some data concerning the present yield in different parts of the world.

Let us consider wheat, the world's most important cultivated plant. In Table IV the yields of the highly advanced agricultural nations of north-west Europe top the list. Notice that not only those European countries have been included but also a number of non-European, principally those with very extensive wheat cultivation. Included are figures for the continents and the world.

Table IV

Yield of wheat in tons per acre
Averages for the period 1948/1949-1953, and the year 1962/63

	1948/1949-1952/1953	1962/1963
Netherlands	1.46	1.82
Denmark	1.46	1.67
Belgium	1.28	1.58
United Kingdom	1.09	1.74
Switzerland	1.06	1.52
West Germany	1.05	1.29
East Germany	1.05	1.24
Ireland	0.91	1.38
Sweden	0.84	1.11
France	0.73	1.23
Austria	0.68	1.64
Europe	0.59	0.84
Canada	0.51	0.57
United States	0.45	0.68
Argentine	0.46	0.58
India	0.27	0.36
Japan	0.74	1.02
Turkey	0.40	0.43
China, Taiwan	0.28	0.84
United Arab Republic	0.74	1.04
Australia	0.45	0.50
New Zealand	1.09	1.20

	1948/1949-1952/1953	1962/1963
North America	0.46	0.64
South America	0.43	0.53
Asia	0.33	0.38
Africa	0.28	0.36
Oceania	0.45	0.51
The World	0.40	0.50

The Netherlands is in the lead, closely followed by a number of West-European countries. In most of the south- and east-European countries the yields are considerably lower. This accounts for the fact that the average for Europe is less than half of those given for the leading countries. Of the non-European countries New Zealand, Japan and the United Arab Republic are foremost, whereas the large cereal producing countries are low down on the scale.

Obviously the sequence can alter a little from one year to another in accordance with climatic changes, but on the whole the figures give a reasonable picture of the situation, but it should be noted that the average yield of wheat over the entire world is less than a third of that in the Netherlands.

However, the maximum wheat output is not by any means reached by even the highest yields per acre listed above. For instance, in the south of Sweden, Scania, two tons per acre have often been harvested, and under very favourable conditions even this can be considerably increased. In 1943 it was reported that two new strains had yielded up to 3 tons per acre. Moreover, yields per acre of the same order as the last mentioned have been obtained from irrigated areas in North America.

Even if the reservation has to be made that north-western Europe has very favourable conditions for high yields, it is hard to avoid the conclusion that, since the highest output of wheat is nearly ten times as large as India's average for 1960/1961, there must be considerable room for improvement in the yield in many countries. It should be remembered, too, that the Indian yield was by no means the lowest; many other countries show lower figures.

Similar conditions prevail with regard to other crops. For example, for 1962/1963 the average rye yield per acre in Switzerland was 1.59 tons, and in the Netherlands 1.27 tons, whereas the global average was 0.49. The average barley yield, for the same year in the Netherlands was 1.72 tons per acre, in Denmark 1.59, and in Belgium 1.56, while the world-average was 0.62. During the same year oats gave an average yield in the Netherlands of 1.56 tons per acre, in Denmark the figure was 1.48, in the United Kingdom 1.37, while the world-average was 0.58. In the year 1962/63 the average rice output was in Spain 2.49, in Australia (in a relatively small area) 2.43, in the United Arab Republic (Egypt) 2.34 tons, in Japan 2.11, and in Italy 2.20, whilst the world-average was 0.80 ton per acre. In India, which has a larger rice-area than any other country, the average yield was only 0.55 tons per acre, and in some African countries no more than 0.32. In 1959/1960 cane sugar gave an average yield of 7.9 tons of sugar per acre in Hawaii; in Indonesia the average yield was 3.9, in Fiji 3.3, and in Cuba 1.4. No figure for the world-average is available.

We could examine the yields of one cultivated plant after another and always find that the world average would lie well below what is obtained in those countries where cultivation is most advanced. And even in these countries there is a long way to go before the maximum output is reached.

This question therefore arises: is it possible to increase the yield, not only in those countries where it is still far too low, but also in those where it is relatively high? My answer is: yes. The ways and means to do so are many.

Let us now consider some of them.

The Regulation of Water

In order that the measures to be discussed in the sections that follow should be successful, the condition of the soil must be such that the plants can take up the nutrition they need. Here, undoubtedly, water supply and its control are of prime importance. Certainly all growth factors are equally important – a deficiency or surplus of, for instance, copper, an essential trace element, *can* be just as decisive to the yield as a deficiency or surplus of water; and yet, in practice, an adequate water supply is nevertheless decisive in most cases. It is just as important to drain the soil of an overabundance of water as to supplement a deficiency of water by irrigation. Both measures *can* be necessary and often are imperative on one and the same piece of land.

There are no satisfactory statistics relating to the total need for drainage and irrigation. But let us for a moment consider the conditions prevailing in a technically advanced country like Sweden. As yet only a third of the land under cultivation is drained. It is well known that those districts where drainage is most thorough are also the most fertile. It is also well known that much improvement is possible in those agricultural regions where drainage is unsatisfactory. The same is true for many countries with a highly developed agriculture.

Enormous tracts of the earth's cultivated area could yield much larger crops than they do at present if they were well drained. Even more extensive are those regions where the yields could be many times greater with adequate irrigation. It is true that many of these regions are so situated that irrigation cannot be effected. Nevertheless, those in which it is possible are of such dimensions that an improved water supply would ensure a very considerable contribution to the world's supply of foodstuffs.

Several of the possibilities of increasing the cultivated area mentioned in a previous section are dependent upon the supply of water.

Here we could add that in India, Thailand and Ceylon only 5 per cent of the rivers' waters is utilized for irrigation purposes. In Burma no more than 1 per cent.

According to an estimate made by FAO, India's water resources should be sufficient to irrigate 650 million acres, and Pakistan's annually cultivated area could be increased by as much as 50 to 70 per cent if the irrigation potential were exploited. It is estimated that there is enough water to irrigate in the United States a further 15,000,000 acres, and in Latin America a further 12,500,000 acres. Larger still are the areas which can be irrigated in China. During the period 1952 to 1960 the effectively irrigated area in that country was enlarged from 52.5 to 140 million acres, and it is estimated that during future decades the irrigated area can be increased to 250 million acres.

It may be claimed without exaggeration that, of the nine points mentioned (on p. 56), the control of water supplies, which means in most cases an improvement in irrigation, is by far the most important. In fact, in most of the earth's less developed, over-populated regions the production of foodstuffs is hampered by lack of water, and a satisfactory solution of the problem of water supply is an absolute condition for progressive development, not only in agriculture but for the whole standard of the community. In arid regions it may be possible to increase the yield per unit area by 50, or even 100 per cent, by such measures as tillage, fertilizing plant breeding and so on (we shall discuss these below), but irrigation can bring about an increase of several hundred, even a thousand per cent. A few examples taken from my experiences during a journey round the world may serve to illustrate this statement.

In Hawaii, where the sugar cane plantations are well irrigated, the sugar yield is on the average more than twice as high as that of Fiji, where there is no irrigation. The enormous sums invested in the irrigation systems of Hawaii – reservoirs, deep wells with powerful pumps, and an elaborate network of canals and conduits for the distribution of the water over the fields – have laid the foundations of a high standard of living. For it is these which have made possible the use of the sunlight which floods the country all the year round.

In a few arid regions of New Zealand practical experiments have shown that it is possible, by means of irrigation to increase the yield from pasture-land by about 1,000 per cent.

In large parts of India, that is, in the semi-arid districts, the wheat yield is less than a fourth of that in some countries in north-west Europe. If the monsoon fails the result is a total crop failure, and if it is too heavy then large areas of cultivated land may be destroyed by water erosion. Generally, however, it is the lack of water which is the limiting factor. Drought forces the peasants to be idle for long periods of time. When I was there – it was during such a dry spell – the wheat was sparse and weak, and the rice fields had dried out. There was practically nothing for the people to do. With an abundant supply of water it would be possible to have two or three crops per year, each one as large as any obtained in north-west Europe – all in all ten times more than at present. It is water that will increase production, promote work, food and health!

Water can be obtained in a variety of ways: by diverting the waters of rivers from more highly situated, rainy districts; by building reservoirs to catch and retain the monsoon rains, so that both soil erosion and flooding can be prevented; by drilling for water to great depths. However, any work of this kind inevitably calls for vast capital investment, which at present is far from adequate.

In India I also saw the works which were then being carried out in the valley of the river Damodar, which flows into the delta of the Ganges south of Calcutta. These works have been financed by loans from the International Bank for Reconstruction and Development. A large number of reservoirs have been built to retain the rain that falls during the wet season. Power stations have been built at four great dams, and a number of smaller reservoirs constructed to regulate the flow of water. From these dams the water is led out to the neighbouring terraced fields. A large quantity of water is harnessed to fulfil two tasks, first to provide electric power, and then to water the crops.

Two other great projects in India may also be mentioned. Along the river Sone 1,025,000 acres of cultivated soil will be irrigated by a system which is being extended and improved, a project made possible by a loan of fifteen million dollars from IDA. In the lower part of the Indus Valley 312,500 acres of arable land which is flooded at times and damaged by salinity is to be put into order and protected. This project is being financed by a loan of eighteen million dollars from IDA.

The Dez dam project in Iran, costing 75 million dollars, will solve the irrigation problem for 352,500 acres of cultivated soil, and at full capacity will provide 520,000 kWs. of electric power.

I also saw large irrigation constructions in Israel. From north of the Jordan Valley water is led in three large pipes to the arid districts of southern Israel. In 1952, in the partial desert of the Negev region, boreholes were sunk to a depth of 600 metres in search of water; these efforts met with success. The importance of water in that region can easily be understood when it is realized that good irrigation makes it possible to have twelve yields of lucerne per year (7.2 tons of dry fodder per acre, or, three times the north-west-European average). In this case no per cent calculation is possible, because without water there is no crop harvest at all.

Another much larger project, which as yet has not gone

beyond the discussion stage, is to make use of all the water in the River Jordan for irrigation purposes – partly in Israel and partly in Jordan. In order that the Dead Sea should not be left dry, water would be led through a canal and tunnel from the Mediterranean. As a result, a drop of 393 metre at the edge of the Dead Sea would come about, making possible the construction of a water power plant with tremendous capacity. In this way the salt content in the Dead Sea would increase and the extraction of salts – for instance, potassium salt – would be greater than it is now. The project is both technically and economically feasible, but up to now it has been prevented by the political disputes between the two countries.

Larger still would be the gain in cultivable area if the Sahara, or at least large portions of it, could be made agriculturally productive. It is known that under this desert there are very large reservoirs of water which could be brought to the surface. The water under the Sahara is located at great depth within a very wide area south of the Atlas mountains. By a great number of borings (for oil) it has been possible to map the extent of the enormous 'Sahara-lake'. It extends from the boundary between Algeria and Morocco in the West to the district south of Tripoli in the East. It is 960 miles long and 180 miles broad and it has been estimated to contain 50,000 milliards of cubic metre water in two layers, the lower one at a depth of 1,700 metre. If these enormous resources could be utilized in a rational way according to methods elaborated in Israel very large parts of the Sahara could be transformed into productive land. It should be emphasized, however, that out of water resources of this kind in the long run only a quantity corresponding to the annual supply can be utilized each year.

According to a French project the energy of the sun could be used to pump up the water in order to make a large area in Algeria productive.

A similar project, though on a smaller scale, has been drawn up in Tunisia to transform the country's desert regions into fertile land. (The Romans had constructed a vast irrigation system there – and the ruins of their aqueducts still stand.)

Questions concerning water supply – a factor to which greater attention has probably been paid in Israel than in any other country – became the dominating subject of the discussions on the problems of agriculture during the Rehovoth conference.

Several of the aforementioned projects presuppose a modern technique of boring which has already been applied with great success in Israel. The results which have been achieved with these modern methods, together with purposeful geological and hydrological research, were presented by Professor L. Picard of the University in Jerusalem.[6] Among other things he said that, from 1950 to 1958, more than 1,000 wells had been drilled, creating an increase in the water resources of 500 million cubic metres per year. In 1959 the figure was 900 million cubic metres. Many of the bore holes go down to a depth of 800 to 1,000 metres.

However, water is needed not only for agriculture but also for human communities and industries. It is therefore necessary, not only to plan the management of water carefully, but also to pay greater attention to the possibilities of conserving it – for instance, by preventing evaporation occurring from irrigated land and from open water surfaces (cf. p. 80) and by adopting appropriate measures to make it possible to use a given quantity of water several times. Professor Lowdermilk gave the following very good example of this (c.f. p. 63). In a catchment area of about 20,000 acres in San Dimas, Southern California, hydrological investigations were instituted in order to ascertain how one should set about obtaining the greatest possible part of the rain for the irrigation of fertile lands. As a result not one cubic metre of water now flows out into the sea. The entire volume of rain is used effectively.

There is one other possible way of obtaining water for arid regions, namely, by desalination of sea-water. Scientists have been working for a long time on the different methods which have been devised in accordance with the three principles: evaporation, electrolysis and freezing.[7] Of these only distillation is, as yet, being used on a small scale for the production of fresh water, for industrial purposes, for household purposes. It is, of course, used on all ocean-going steamships today. Electrolysis has so far only been used on a small scale; it appears to be

[6] *Ground Water Exploration in Israel.*

[7] G. Dumesnil, *Desalination of Sea-Water.* I. Vilentchuk, *Sea-Water distillation as a major addition to the Natural Water Resources.* A. Zarchin, *Desalination of Sea-Water by Freezing.*

the method best suited for water with a salt content smaller than that of sea-water. On the other hand, freezing by Zarchin's method has already been employed on a fairly big scale: for instance, in the Israeli town of Eilat on the Red Sea about 1,000 cubic metres of desalinated water per day are being distributed through the town's mains.

At present one would have to reckon that, under reasonably favourable conditions, the cost of desalinated water would be in the region of eightpence per cubic metre. For irrigational purposes that is rather high. However, intensive research is being carried on to improve and cheapen methods of desalinating seawater. A link in the efforts to reduce the cost of fresh water is the project which is now being developed in Israel, in co-operation with the United States, for generating electric power and desalinating sea-water by atomic energy. For generating electric power steam of high temperature and under high pressure is required and for evaporation steam of lower temperature and under lower pressure. By combining the two purposes the steam is first used for generating electric power and then for desalination. The cost of fresh water production is then reduced. The cost can be decreased by enlarging the combined establishment; according to a press report last year, an agreement had been reached between the United States and the Soviet Union to develop a project of this kind with a much bigger capacity than the Israeli project. The results of further research will undoubtedly be of vital significance for many countries in the arid regions of the world. In order that desalination methods may be used in a practical way to solve the agricultural water problem it is essential, of course, that there is access to cheap energy. The great progress which has been made in recent years in the endeavour to harness the energy of the sun, gives rise to the hope that the time is not far off when sunlight will be the force enabling us to supply the arid, unproductive regions of the earth with life-giving water.

Salt water can be used for irrigation purposes when the salt content is low. This emerges from the research which Dr Hugo Boyko has undertaken at the Negev Institute during recent years. He gave an account of this work at a meeting during the Rehovoth Conference and later in New York. However, his method can only be applied to very porous sandy soils, for example, fields of drift sand with a very low content of clay material. Upon more clayey soils the surface layer gradually becomes enriched with salt with the result that in the long run crop growing is made impossible. But within the arid and semi-arid regions there are, as is well known, enormous areas with sandy soils. Within the dry regions Boyko's method may, therefore, be of very great importance. The water has to be used very freely so that the salts are washed down to those layers of soil which lie below that in which the roots of the plants are spread. By mixing fresh water with salt water it is possible to reduce quite considerably the cost of irrigating more or less salt-tolerant plants.

There are great differences between plant species in their tolerance of salt. Several cultivated plants can be watered with water containing 0.6 per cent salt (i.e. the salt content of the Baltic); others endure much less, while still others, for instance sugar beet, barley and lucerne, endure more.[8] The higher the salt-content, the more limited the variety of plants that can be cultivated. Boyko was of the opinion that it might be possible, by plant breeding, to produce plant varieties which would be more tolerant of salt than those cultivated hitherto.*

However, irrigation in arid regions can create a problem which the farmer in a humid district never has to face, namely, salt-accumulation in the surface layer of the soil. Due to strong evaporation the water in the soil moves upwards and the salts, dissolved in water, collect in the surface layer and can reach such a concentration that the plants are damaged. This process, salination, is hastened when the level of the water table is raised as a result of irrigation. A rising water table can lead to rapid salinity not only of the irrigated area but also in neighbouring fields. Irrigation must therefore be effected so that salination is counteracted; this means, among other things, that it must be combined with good drainage. In order to avoid the loss of water through evaporation from the soil surface and the salination of the top soil layer an Israeli engineer has developed the method of subterranean irrigation, that is the water is supplied through a richly ramified system of plastic tubes, situated in the soil layer where the main part of the plant roots occur. Consequently, the soil surface remains dry. Salinity is one of the

[8] Boyko 1964.

* See Note on page 139.

most pressing problems to which research stations in the arid regions are devoting their attention. It has been solved by the method devised by Boyko, that is, it must be possible to give, now and then, such an abundance of water that the salts are washed down into the deeper layers of soil. Since salinity is combined with a high content of natrium, it can become necessary to treat the soil with gypsum – up to four tons per acre.[9]

Water can also be obtained by creating rain. As is known, rain is formed by the condensation of water-vapour round very fine, solid particles. Drop-formation in the clouds has been very intensively studied by the meteorologists, and American scientists – later also Australian – have succeeded in creating rain by 'seeding' super-cool clouds with carbon-dioxide snow from aircraft. Later on it was found that silver iodide, which has the same structure as ice crystals, and fine salt-powder had the same effect. It follows that rain is easily formed in coastal districts round the fine salt-powder coming in from the sea.

A group of scientists from UNESCO, who at the beginning of the fifties studied these problems in the Punjab in Pakistan, found that the air there, because of the long distance from the sea, contained extremely little salt. It was therefore natural that the possibilities of producing rain by increasing the quantity of salt in the air should be investigated. A very simple method was devised – salt was pulverized very finely and the powder was blown out into the air by simple handsprays.

In 1954 a decisive test was made. For two months salt-powder was sprayed out over two large regions in the Punjab. The result was that 50 per cent more rain fell in the 'salted' areas than in the neighbouring 'untreated' areas. In relation to the achieved result the method was cheap and it was said that Pakistan would use it on a larger scale to induce rain over the country's arid region.[10]

Experiments have been made in other arid regions to produce 'artificial' rain, above all in Australia and Arizona. During the Rehovoth Conference, the Principal of the Australian Institute of Radiophysics, Dr E. G. Bowen, gave an account of the experiments made in Australia.[11] In a number of experiments, under-

[9] Verhoeven, 1961.

[10] To what extent this has been done I don't know.

[11] 'Climate Control to meet Problems of Aridity (Rainmaking)'.

taken over periods of three to five years, super-cool clouds were sprayed with silver iodide from aeroplanes. There were varying results. In two experiments the rainfall increased by 10 to 30 per cent, in one the result had been negative, and in two other districts the results seemed to have been favourable. Dr Bowen was of the opinion that there are extensive regions in which the silver iodide treatment can be successfully applied.

Similar results had been obtained from experiments made in Arizona. These were described at Rehovoth by Dr Louis J. Battan of the Institute of Atmospheric Physics at Arizona University.[12] However, the increases of rainfall were less than those obtained in Australia and not statistically reliable.

Even if it is technically possible to induce rain by various methods, it must nevertheless be admitted that the possibilities of reaching a final solution of the problem of water supply in such ways are limited. The different layers of the atmosphere are in constant movement and if rain is induced over one region, it may well be that the natural rainfall over another is reduced. Thus an area which has previously been relatively arid could become even more so. It is therefore not unlikely that such rain-producing methods could give rise to conflicts between different countries.

In short, I think it would be wise not to expect too much from these methods, but rather work with the water which is already on or just below the earth's surface – in other words, concentrate on the irrigation methods discussed earlier on.

There is, however, another possibility which deserves our attention. In 1961 Professor Tor Bergeron, the Swedish meteorologist who has given considerable thought to the problem of rain-formation, discussed with Dr Bowen the possibility of reducing the formation of rain in those regions where precipitation is too abundant. By this means the quantity of rain in neighbouring regions less well-watered, might be increased. One way in which this could be done, they thought, would be by increasing the number of condensation nuclei in the rainy district to such an extent that the 'cloud drops' become far too numerous and, as a consequence, too small to fall as rain. But later, when the air is above rain-deficient areas the moisture might possibly fall out as rain.

If such a method were successful, a twofold result could be obtained. In the first place, the violent precipitation in the tropical rain-forest belt could be diminished, and erosion reduced at the same time; while in the second place, the rainfall could be increased in the more arid regions at those periods when it is most needed. So far no practical experiments have been carried out to test this method, but the idea has attracted interest among the scientists working on the problem in the arid regions.

It would have valuable results from quite another point of view if it were possible to prevent too violent storms of rain and hail. In the United States and Europe about £20,000,000 are spent annually on insurance premiums against hail.[12] Research into the possibilities of avoiding such storm damage is a matter of some urgency.

In many arid and semi-arid regions shifting sand presents a problem difficult to overcome. Not only is the sand worthless from an agricultural point of view, but it also constitutes a constant threat to neighbouring zones. It shifts across cultivated areas, blocks water-courses, and buries buildings, roads and railways. History provides many examples of extensive civilized tracts which have been destroyed by shifting sand.

In Israel the problem of binding the sand and transforming areas of shifting sand into fertile fields has been solved. The significance of this achievement becomes clear when it is known that in Israel alone there are 250,000 acres of shifting sand, 162,500 of which are in Negev. By irrigation, by planting sand-binding plants, by fertilizers, and so on, four years have seen the successful transformation of the light yellow drifting sand into a fertile, dark brown arable land which has given an annual yield of fifteen to twenty tons of dry fodder per hectare. The Israeli scientist who has led these undertakings, Professor Ravikovitch, concludes his report with the following words: 'Fields of shifting sand can be transformed into arable land in a relatively short period of time.'

It may be added that during recent years great progress has been made in Israel in the endeavour to utilize the arid regions. Apparatus has been successfully constructed which makes possible the use of a considerable amount of the sun's energy (see

[12] Haslett, 1963.

below under the 'Supply of Energy') to produce steam-power at approximately the same price as that obtained by conventional fuel. This enables the production, in the arid regions, of the sources of energy which are required for irrigation networks, and also for total or partial desalination of water with a too high salt content.

To summarize, it can be said that the regulation of the water supply of cultivated lands is of fundamental importance to secure a satisfying return of other measures of cultivation, such as liming, fertilizing and tilling, and to obtain the greatest possible yield of the varieties of cultivated plants obtained by breeding. Water regulation is, therefore, one of the most important conditions of increasing the yield from cultivated soil.

To realize this condition landscape-planning is required in many countries, as Professor Lowdermilk has emphasized. Such planning must be based on a comprehensive investigation and mapping of the natural resources, above all, of water and types of soil; on the use to which soils are put; on the deterioration of soil through wind and rain, and so on.

These ideas might be realized in the near future, since FAO, as a part of its campaign for 'Freedom from Hunger', has included a detailed mapping of the soils of the world. Because of the differences which exist in the reaction of various types of soil to different methods of treatment, these maps can be of a great value in determining the use to which the earth's hitherto uncultivated areas may be put. Such maps will fill the need, mentioned by Lowdermilk, for methodic planning.

Technical assistance to less developed countries in the development of their agriculture and in water management should, in Lowdermilk's opinion, be given in the form of demonstration farm complexes of five to ten thousand hectares each. These should be situated in districts which are not too badly damaged and which could be rehabilitated in a relatively short time. In these areas, agro-industrial pilot projects, with the necessary irrigation and rationalized cultivation methods, would then be undertaken.

A fascinating technical problem concerns the reduction of evaporation from exposed water surfaces. In arid regions the volume of water which can be lost through evaporation – for example, from a reservoir – can be larger than that used for consumption. The water situation would obviously be much improved if this loss could be prevented. Very promising results have been obtained by the application of monomolecular films of cetylalcohol on the surface of the water. However, before this method can be used on an economic scale the technique must be further developed. The possibility of avoiding loss through surface evaporation by using subterranean irrigation through plastic tubes has already been mentioned. A plastic (or asphalt) product has also been used in another way. The Esso Company has developed a preparation, Encap, which is sprayed on the soil surface, where it forms a thin, coherent, but somewhat porous coat which considerably reduces evaporation.

It is doubtless clear from the previous arguments that the potential foodstuff production of many of the earth's arid regions is very considerable – in fact, enormous. UNESCO's magnificent programme, Arid Zone Research, which was started in order to promote research concerning the very large arid regions (arid and semi-arid regions comprise more than one-third of the earth's surface) was concluded in 1962 after six years intensive work. During this period both national and international activity was very considerably intensified, and basic research as well as the training of research workers, received a powerful stimulus. The many valuable results can be of fundamental significance for the future. That the project has been ended does not mean that research into the problems of the arid zones ceases; on the contrary, it is being continued as a part of UNESCO's regular working programme. It is being pursued, in fact, on a larger scale than before by direct and practical measures made possible by UN's Expanded Programme of Technical Assistance (EPTA) and UN's Special Fund for International Development (SF). In a UNESCO publication (1962) the geographer, Professor Gilbert F. White, Chicago University, describes the present situation and discusses some of the results obtained.

Liming and Fertilizing

To produce a satisfactory yield from cultivated plants the soil must contain lime and a large number of nutritive substances in reasonable quantities. For normal development the plant needs relatively large quantities of some substances and very small quantities of others. But both are equally essential.

Lime is a nutritive substance, but it is also needed to create, or maintain, a soil reaction suitable for cultivated plants and for many of the soil's micro-organisms. Soil reaction must meet the varying demands of different types of cultivated plants. Some develop best in slightly acid soils, others thrive best where the reaction is neutral or slightly alkaline. Others again are relatively indifferent. Every year large quantities of lime are removed from the soil by the plants themselves and by the water which runs off the land; where the rainfall is high the greater the loss of lime. The soil must therefore be supplied with lime in varying quantity from time to time. In most countries the lime supplies are quite sufficient for agricultural needs.

Usually the amount of nutritive substances in cultivated soils and accessible to the growing plants is not sufficient to meet their needs. On those farms where animal husbandry is combined with crop production, part of the nutritive substances are annually brought back into the soil in the form of animal manure; but this contribution is inadequate and has to be supplemented by chemical fertilizers to produce a satisfactory yield.

Of particular importance is the application of artificial fertilizers in districts where animal manure is used as fuel, no wood being available because the forests have long since been laid waste. Many million tons of animal manure are wasted in this way every year.

Since the end of the Second World War the world consumption of artificial fertilizers has been about doubled. But the increase has taken place on the whole in those countries where the yields were already high. In the under-developed countries the increase is insignificant; indeed, in many countries no artificial fertilizers are used at all, and there is a great need to spread information concerning them so that this vital factor may be brought into full use.

During the Rehovoth Conference great attention was paid to the problems of fertilizing. Dr Frank W. Parker, FAO, in a lecture on the role of artificial fertilizers,[13] emphasized that it was not least in the new states that their use should be substantially increased. 'Hungry plants make hungry people,' he said. Without artificial fertilizers most soils can yield 800-1,400 kgs. of grain per hectare. A hundred years ago the crops in Europe and Japan were of about that size, and in most less advanced countries they are roughly of this size today. Together with other measures, the use of artificial fertilizers has doubled or trebled the crops of many countries. In large parts of Asia, Africa and South America it is essential to use artificial fertilizers generously if the crops are to be of a reasonable size. In the next few decades it will be imperative to attend to the provision of plant nutrition in these parts of the world; among other things, by building factories for the manufacture of fertilizers.

Baade estimates that the ideal consumption per acre of nitrogen, phosphates and potash is half a hundredweight of each, that is, about ten times as much as at present. It should be added that in many districts in Western Europe the quantities of fertilizers are higher than those mentioned.

With a view to increasing production and the use of artificial fertilizers in the under-developed countries FAO has worked out a fertilizer programme as a part of its 'Freedom from Hunger' campaign.

The known resources of phosphorus and potassium are sufficient to allow a much more generous application of these nutrients on those soils where they are already being used, and to extend their application over much wider areas. So far as nitrogen is concerned, the supply is literally inexhaustible. Up to the present, artificial fertilizers are mainly used in Europe, North America and Japan, which together use about 90 per cent of what is manufactured. A rational use of fertilizers on the cultivated lands of the earth could bring about a very considerable rise in production, but it is hardly possible to give even an approximate estimate of the potential increase. It is no exaggeration, however, to claim that in those regions where the water supply is good a rational use of animal manure and artificial fertilizers, together with the growing of the nitrogen fixing

[13] The Role of Chemical Fertilizers in Increasing Agricultural Production in Under-developed Countries.

leguminous plants, could increase the yield by 50 to 100 per cent, indeed, in large regions of the world, by as much as 100 to 200 per cent. But even if the possibilities of development are great, it must, of course, be remembered that it will take time before an even more or less acceptable state of affairs is reached.

Potassium, phosphorus and nitrogen are the three elements of which comparatively large quantities are needed, and which farmers in the developed countries have used, to a greater or lesser degree, for a long time. But even if fertilizing with these substances is carried out adequately, the harvest can fail completely if the soil is deficient in one of the micro-elements, or trace elements, essential for normal plant growth. Among these essential substances, of which, as the term implies, only very minute quantities are needed for healthy development, are to be found copper, zinc, manganese, cobalt, molybdenum and boron.

At the end of the nineteen-twenties it was discovered that, among other things, copper was essential if oats were to grow normally. Without copper the grain did not develop, but with copper the yield could be 1.6-2 tons per acre. Later it was found that other cultivated plants also need copper in small quantities. By putting these results into practice the harvests could be appreciably increased in some districts in the Netherlands, Germany, Denmark and Sweden where the soil was deficient in copper.

A vast region in South Australia, with the significant name '90 miles desert', was formerly a waste land of steppe and desert where 10 to 15 hectares (25 to 37 acres) were needed to maintain one sheep. Many attempts to cultivate this region had failed in spite of the fact that the rainfall was not particularly low. At the beginning of the 1940s it became clear that one of the reasons for the low yield was a deficiency of copper and zinc in the soil. It was soon found that the 'desert's' 3.75 million acres could be made productive by fertilizing with copper and zinc sulphate at a cost of about 6s 6d per acre. Many farmers who have made practical use of the results of scientific research have been able to increase the number of sheep on their lands very considerably. Wide areas of the former 'desert' are now cultivated not only as pasture but also as arable land.

There are similar districts where a deficiency of molybdenum has prevented a good yield and where, by the application of a very small dressing – 200 grammes per acre – the number of grazing sheep has been increased from one to 100 per 25 acres, and where the value of the land has gone up from about 30s to £55 per hectare. Science has conquered the desert sands. The situation is similar with regard to many other trace elements.

Cultivation and Mechanization

Among the measures which can contribute towards an increase in the earth's supplies of foodstuffs is the increase in mechanization and motorization. This does not mean that these measures, as such, increase yields. It is quite obvious that a good state of cultivation can also be achieved with animal-drawn implements and that the crop can be harvested by hand. There are, however, instances where the motor-plough has been absolutely essential to enable the farmer to cultivate certain areas. For instance, in certain parts of India large, fertile areas were infested with the wild sugar cane (*Saccharum spontaneum*), and its tough rootstocks, inches thick, precluded cultivation by means of the primitive, animal drawn implements.

It is also essential to be able to cultivate the soil and harvest the crop at the right time, and with the little available labour in most advanced countries, mechanizaton and motorization are absolutely necessary if favourable results are to be obtained. In this sense, both mechanization and motorization may be regarded as playing a very important role in the endeavour to increase yield.

An important aspect of the question of mechanization, or rather, motorization, has already been touched upon. For a very large part of the products of the earth is used as fodder for draught animals; in other words, for the production of animal-power. Extensive motorization would free large areas for the production of food which at present are used to produce fodder. Sweden provides a good example of this. Since the last war some 750,000 to 1,000,000 acres have been made available for the production of foodstuffs which were previously used for producing animal fodder (cf. p. 52). In this particular case fuel (for the tractors) has been substituted for oats (for the horses).

With regard to the mechanization of agricultural methods as practised in many of the poorer countries, consideration must be taken to the fact that development in food production must go hand in hand with advancement in other branches of industry and commerce. Should the former progress too rapidly, the consequence can be that masses of people become unemployed.

Plant Breeding

By improving the strains of cultivated plants, very considerable increases in yield can be achieved.

The varieties of wheat now cultivated in south Sweden give 50 to 60 per cent higher yields than the old, unselected homegrown varieties. The progress made in very recent years lends support to the assumption that maximum outputs have not yet been reached. When the cultivation of sugar beet was begun in the first decade of the nineteenth century, the beet had a sugar-content of 5 per cent whereas nowadays the content is nearly four times as great. Another telling example can be seen in the hybrid-maize varieties which were first introduced in the United States; by this hybridization the yield was increased by about 25 per cent. The yield obtained by the development in Java of the new variety of sugar cane, POJ 28, was increased by the same degree.

Similar results have been obtained in other parts of the world. However, by far the majority of varieties cultivated today have not yet been improved by the plant breeders. Tasks of prime importance are here waiting to be performed.

One of the most important of these tasks is to produce better strains of rice, which is, next to wheat, the world's most important food, cultivated on about 300 million acres. Several countries are presently developing improved strains of rice, until now only Japan has obtained noticeable results. Since 1949, under the leadership of FAO, co-operative research is being carried out with the purpose of transferring the good yield-capacity of the Japanese rice to those varieties which are cultivated in countries with sub-tropical climate.

Another important task is to improve the millet varieties, since they provide a considerable part of the food grown in large areas of Asia and Africa. Very valuable results may soon be forthcoming, although it will probably be some time before they can be put into cultivation on a sufficiently large scale.

It is occasionally claimed that the quality of grain has deteriorated as a result of plant breeding, and that similar breeding could cause deterioration in the quality of other cultivated plants. It should, therefore, be mentioned that when wheat breeding began in Sweden, attention was indeed directed to a quantitative increase which, among other things, resulted in the newer strains having a lower protein content than the older strains. This was not then of such vital importance, as the vast quantities of wheat which were being imported at the time had a high protein content. However, when Sweden approached the state of being self-supporting so far as wheat was concerned, attention was directed to the development of strains which combined high yield with high quality. Our plant-breeders were very successful, and modern improved varieties have certainly no less nutritive value than the old varieties. Similar work is being done with all cultivated plants. In other words, research is intent upon improvement not only of yield but also of quality. It may well be possible also to change the composition of the vegetable proteins.

Plant breeding does not confine itself to the improvement of plants already under cultivation; it may also give us new types of plants. An example of this is the rye-wheat which may be placed on the market in the near future and could increase the earth's productivity in many lands.

In some of the under-developed countries work has been set in train to improve strains. But the personnel available and able to carry out such research is quite insufficient. The step taken by the Central Committee of the Swedish Technical Assistance Organisation, which in 1960 arranged for a course on plant breeding at Svalöv and Lund in which representatives from several countries participated, was most important. This course was received with great interest and the work is being continued by the Swedish Agency for International Assistance (now the Swedish International Development Agency) in co-operation with FAO.

Increased Yields by Preventing Flower-Formation

Science has shown many other ways of increasing yield per area unit besides those already described. Methods have been developed in Hawaii to increase the yield of sugar cane by preventing the plant from flowering. Quite a large amount of sugar is consumed by the plant in the formation of the flower and fruit. Those plants of which we utilize the vegetative parts – root, stem or leaf – in which sugar or starch are stored, can give a higher yield if the development of the reproductive parts – blossom and fruit – can be prevented. This can be done in several different ways. Common to all methods is the checking of the process of photosynthesis at that stage in the growth of the plant when flower-formation is about to begin. The treatment in question must, therefore, be begun at or just before this moment. Experiments have shown that sugar cane in which the formation of the flower has been prevented gives an increased yield of 10 to 20 per cent, or two to four tons of sugar per hectare.

One method which has been experimentally tested, but which can only be applied in the treatment of 'short-day' plants (to which the sugar cane belongs) is to illuminate the cultivated fields during the dark hours of the night. Ten hours of illumination, spread out over ten periods of one hour each per night, have proved sufficient. For this lighting low-flying aircraft equipped with powerful searchlights have been used. Another method is to expose the cultivated fields to drought by halting irrigation. A short break in the supply of water does not cause any lasting damage but does completely prevent the formation of flowers. A third method is to treat the crops with certain chemical preparations. Several have been tested which have either reduced or completely prevented flower-formation.

Of these three methods, the short period of 'drying out' is that which has been used most.

Such methods of increasing yields could, perhaps, be used on other cultivated plants too, where the yield consists of the vegetative parts which, as has already been mentioned, are prevented from growing as fully as they might because of the development of the reproductive organs.

The Control of Plant Diseases, Insect Pests and Weeds

The control of plant diseases, insect pests and weeds, constitutes a significant group of measures in the endeavour to increase food production.

The losses caused annually by parasites of various kinds

involve enormous sums. At the Second International Congress for Plant Protection, held in London in 1949, the president said in his speech of welcome that the damage caused by plant diseases and insect pests to growing crops and stored products had been conservatively estimated at 30 per cent, and this did not include the losses caused by the 'competing' weeds. He said that FAO had estimated that the total damage to the world's production of cereals alone, as a result of disease and pest, had caused an annual loss of 65 million tons. If this loss could be reduced by even as much as 10 per cent, it would already mean a very considerable improvement in the world's food supply. Chemical science and industry have provided us with the weapons to reduce these losses appreciably. Indeed, the possibilities in this respect are far from exhausted; on the contrary, we are only at the very beginning of a development which is full of promise.

Different kinds of chemical substances have long been used in the fight against plant diseases. During the past few decades a great number of new preparations have come into the market by which the fight can be conducted with considerable success. Many very good results have been obtained.

Even more tangible are the results which have been achieved by chemical means in the control of insect pests. It is no doubt fairly common knowledge that one of the first insecticides, DDT, has proved very successful.

For quite obvious reasons interest has been directed especially to the fight against the locust, probably the worst parasite of all. There are a great number of locust species, of which one, the so-called Egyptian locust, belonging to the group of swarming locusts, has plagued mankind for thousands of years, especially in the Middle East. The prophet Joel gave a dramatic description of the devastation caused by this pest. In Chapter I we can read: – 'Awake . . . For a nation is come upon my land, strong, and impossible to count; his teeth are the teeth of a lion, and he hath the jaw teeth of a great lion.... Is not the meat cut off before our eyes, yea, joy and gladness from the house of our God?'

From the dry steppe regions, where breeding takes place, the locusts spread in immense swarms over wide areas. According to Vallega, swarms weighing from 50 to 100 tons can be carried by the wind for hundreds of miles. We have a striking example of this from the beginning of the nineteen-fifties when vast areas, stretching from the eastern border of India to the western coast of North Africa were devastated by locusts – one of the largest invasions to have threatened these countries for a long time. Due to the very favourable conditions during the winter of 1948 to 1949, the locusts had become unusually numerous. The result, as already mentioned, was that the swarms spread over many countries in 1950 and 1951. Iran, for example was laid waste by an invasion the like of which had not been experienced for a hundred years. In some regions the earth was covered by a layer of locusts ten inches thick, and where the swarms had passed only the bare soil remained. It is readily understood that the destruction is complete when it is known that a locust can daily consume as much as four times its own weight. Soil erosion is often the result of such devastation.

In Australia conditions seem to have been similar: It was said that in October 1950 several trains had been brought to a dead stop by swarms of locusts in Southern Australia. One train had met with immense swarms, one of which was about six miles long and just as wide.

The fight against the locust is now being led by The Anti-Locust Research Centre, London, in conjunction with FAO. This Centre is conducting intensive research into the biology of the locust, and also into the methods of fighting against it.[14]

One method is to spray DDT, or some other insecticide, from the air upon those areas where the locusts are hatched. The cost of this work, which is carried on in the whole of North Africa, in the Near East as far afield as India, is stated to be many millions of dollars every year. But the money is well spent, since the campaign has succeeded in reducing the number of swarms by several hundred thousand. What this means in terms of economics – and, consequently, also in terms of food-supply – may be illustrated by the following example. Since the war Jordan has been seriously plagued by locusts thirteen times, and each visitation has cost the country, on an average, £200,000.

Locusts appear in other parts of the world than those just mentioned. They are serious pests both in the Soviet Union and

[14] Haskell, 1962.

in South America. In the United States, during the period 1937 to 1949, the damage to crops caused by locusts resulted in a loss in yield amounting to an average of ten million dollars per year, in spite of well-organized measures to combat the trouble.[15]

As artificial irrigation increases, the need to fight the locust becomes more acute. It has become apparent that, whereas in arid regions the female locust lays eggs only once a year, in irrigated regions several generations can be produced annually; the reason for this being that the continuously moist soil produces fresh nourishment in abundant quantities throughout most of the year.

Irrigation has likewise made possible the spread of many other pestilent insects, among them being the malaria mosquito and the tsetse fly.

Apart from campaigns against the locust, the use of other new chemical preparations (insecticides, etc.) is almost exclusively confined to North America and north-west Europe. Their general application over the whole earth would result in a very appreciable increase in food-supplies.

Throughout the ages weeds have been a bane to the farmer. In earlier times the struggle against weeds was mainly waged by various methods of cultivation and by cleaning the seed. Nowadays there are a large number of chemical preparations which are very effective against many tenacious weeds. But it is true of these, too, that they are used more or less exclusively in countries with a highly developed agriculture.

The fact should not be ignored, however, that the use of poisonous chemical substances may also have unwelcome consequence, for example, in that useful organisms are killed. The application of chemical poisons has, on this account, been subjected to severe, though often exaggerated criticism in recent years. Moreover, it is well known that many pestilent insects have developed immunity against these poisons in the course of time. It is therefore important to devise methods of spreading the poisons in such a way that detrimental and undesirable effects are avoided as far as possible. Biological and technical research is working with a sense of urgency to manufacture new preparations with more selective or specific action, or preparations which are quickly decomposed in the soil. The

[15] Uvarov, 1961.

results which have already been obtained with chemical poisons are far too valuable for us to desist from availing ourselves of such an opportunity to increase the world's supplies of foodstuffs.

Over and above the means just discussed, scientific research has also been able to indicate purely biological methods which can be applied in the fight against crop-destroyers. In Australia, for instance, a species of cactus, belonging to the genus *Opuntia*, had been imported as a decorative plant. It spread itself over extensive areas, and in the State of Queensland alone had rendered twenty-five million acres of pasture land practically useless. However, by scattering millions of eggs of the moth *Cactoblastis cactorum*, this 'weed' was brought under control. *Cactoblastis* lays its eggs in the ovule of the cactus and thereby prevents its dissemination. It is then a relatively simple matter to clear the land of the cactus growth. In one small community, entirely dependent on the yield of the sheep pastures, a memorial hall, called 'Cactoblastis Hall', was built in memory of the successful campaign.

A classical example of the successful struggle against an insect pest by biological methods can be seen in Fiji. On the largest island, Viti Levu, it had long been impossible to produce copra, because the cocoa-tree was attacked by an insect, the purple moth *Levuana iridescens*. At the beginning of the nineteen-twenties the moth began to spread to the other islands. This was a severe threat to the most important industry of these islands. In 1924 the government appropriated a large sum of money for entomological research, and after an intense study the scientists found a fly, *Ptychomyia remota*, which on the Malaya peninsula occurred as a parasite upon a moth closely related to the purple moth. This fly (and some other less effective species) was brought to Fiji, multiplied and spread in the cocoa plantations. The flies attacked the purple moth, and after a couple of years the copra industry was able to start again on Viti Levu, and the most important branch of industry on the other islands was saved: an excellent example of successful biological research in the struggle against insect pests.[16]

Another biological method is the spread of insect killing bacteria. *Bacillus thuringiensis*, for example, destroys the larvae

[16] Tothill *et al*, 1930.

of leaf eating insects which devastate many cultivated plants. This bacteria does not kill other organisms, nor does it leave behind any noxious remains; moreover, it can be mass produced on artificial media and be stored and spread like a chemical preparation. It is applied in the form of powder, 10 to 130 lb. per hectare.

At present it is difficult to gauge the significance these and similar biological methods may have in the future. But one thing is certain – a great deal of research is still needed into these measures, both chemical and biological, against weeds and insect pests, if the fight is to be truly successful.

To the biological methods may be added the breeding of disease-resistant varieties. In this field of research very positive results have already been gained. In North America especially, work has proved successful in combating several severe cereal diseases. Perhaps the most interesting example is the production in the United States, after fifty years of patient work, of wheat varieties which are immune to the devastating black rust disease.

Good results have also emerged with other cultivated plants in both America and Europe, and also in Australia and New Zealand.

But man cannot rest on his laurels. . . . Through mutations new strains of disease agents make their appearance, which can attack varieties of plants formerly resistant. Research into the breeding of new disease resistant varieties must not end.

A good example is offered by potato breeding. At the beginning of the nineteen-thirties potato breeders in Germany had succeeded in producing a couple of varieties which were resistant to the severe 'potato disease' caused by the leaf mould *Phytophthora infestans*. However, some years later new strains of *Phytophthora* appeared, which attacked also the previously resistant varieties, and the breeding for resistance had to be continued.

Large quantities of foodstuffs are destroyed by fungi and by insect pests while in storage or in course of transportation. At an FAO convention in 1947 it was estimated that 10 per cent of the world's stock of cereals was lost in this way. Quite naturally the losses are larger in the tropics than in the cooler climates of the northern countries, but even in the latter it is not only perishable foodstuffs which may be destroyed, but grain also. By modern canning, preserving and freezing and the research into related food-preservation problems, it has, however, been made possible to save very considerable quantities of foodstuffs from destruction.

Animal Husbandry

Even if the measures discussed hitherto play the main role in increasing the total quantity of foodstuffs, it should not be forgotten that efficient animal husbandry is of paramount importance. Very considerable quantities of the earth's crops are converted into animal products – in Sweden, for instance, this is estimated to be about 75 per cent. The world average is certainly much less, but very considerable nevertheless. According to figures published by the so-called Cereal Group of FAO, based on information from all countries – apart from China, the Soviet Union and the other East European states – 50 per cent of the total consumption of grain in 1956 was by animals, 40 per cent by human beings, and the remaining 10 per cent was retained for seed, etc. During the following six years the world consumption of grain had increased by 13,000,000 tons, or 3.4 per cent annually; however, the amount of grain used as fodder had increased two and a half times as much as that used for human food. Measures aiming to change this trend are obviously of great urgency.

By improving the efficiency of domestic animals by breeding and by changed method of feeding, it is possible to achieve a more economic result from the conversion of plant products to animal products. This can best be illustrated by figures published by FAO. In the Far East it is estimated that there are 344 million dairy cattle, and that the annual production of milk is 56 million tons, or 410 lb. per cow. In Europe the corresponding figures are 108, 178 and 4,100, respectively – and, it should be remembered, the last mentioned figure lies well below what can be attained.

In most countries with primitive methods of agriculture there is little or no purposeful breeding of domestic animals; in some it has just begun. In all these countries methods of feeding leave much to be desired. Here is an example. I had occasion to visit an experimental farm in the Damodar Valley, where the manager had procured two good pedigree bulls of the local variety. However, it had proved difficult to convince farmers of the necessity of availing themselves of these bulls for breeding purposes, for they thought it entailed unnecessary trouble. In their view, all that one had to do was to let the animals out into the jungle, and, when in heat, the cows would be serviced without further ado. If this is the state of affairs in India, what must it be like in countries where ignorance is more abysmal.

In many cases it is not enough to attempt to improve existing livestock; new cattle must be brought in. Careful and purposeful breeding is then required to adapt the animals to the new environment. In some regions this method has already shown good results.

Considerable and rapid progress should be possible if measures are taken to improve feeding. In all too many districts domestic animals not only receive too little to eat (what applies to human beings also applies to animals), but many get fodder of inferior quality. In a recent report from FAO an account is given of experiments with different methods of feeding. In one of these a litter of pigs was divided into two groups. One of these was fed according to what was generally used at the beginning of this century, the other was fed according to modern methods based on principles founded on scientific research. After two months the pigs in the first group had, on an average, increased in weight by 3.5 kilogrammes (6.7 lb.), in the second group by 49.5 kilogrammes (109 lb.). These and similar results with other domestic animals are founded on intense research concerning the role of different amino acids. These can now be produced artificially and used for improving the fodder.

If we assume (only for the sake of argument) that these and similar results of scientific research could be applied in animal husbandry in the Far East and in other under-developed areas, it would be possible not only to increase the production of animal foodstuffs but also to free for production of human food large areas which are now used for fodder production. An improvement in the standard of livestock would increase food resources considerably because vegetable products would be used more economically.

In tropical regions animal husbandry is faced with special difficulties. In the rainy districts the pasture-vegetation has a high content of fibrous cells and water, but the content of protein is low. In the arid districts the content of dry substance is certainly high, but since the cattle graze all the year round, their fodder is for the most part comprised of dry grass with a high fibre-content. In both cases low production (milk, for instance) is the result. Intensive research in this field is essential for producing suitable forage plants. A number of promising results have already been obtained, but much remains to be done.[17]

Of great importance for the advancement of animal husbandry (and thereby of the standard of living in many countries) is the establishment of modern dairies in milk-producing districts. In this connection the new Swedish methods for ultra-high-temperature (UHT) pasteurization and aseptic milk-packing offer interesting possibilities. The UHT-methods are based on heating the milk to a very high temperature during a very short time (a few seconds). Directly after heat treatment the milk is filled into a sterile Tetra Pak. The product is a bacteria-free milk with taste and nutritional value comparable with pasteurized milk, but keeping for several weeks without refrigeration. Depending upon which packaging material is chosen, a keeping time from one week to several months can be achieved.

The method can also be used for reconstituted milk based on milk powder. This long-life milk is expected to be of great importance, above all in tropical countries where milk keeps only for a few hours. There, an enormous market could be opened up for the milk and this cheap food could be the same important factor in the diet of the peoples of the under-developed countries as it is now in industrial countries.

The control of disease and pests plays a large part in animal husbandry too. In many countries animal diseases are allowed to ravage the herds quite freely. Two examples may be mentioned here. During the Second World War nearly one million cattle died of the plague in Burma; the farmers lost their draught animals, their milk – and the manure for fuel. The result, among other things, was that Burma's total export of rice, usually 3,000,000 tons annually, ceased completely. During the years 1951 to 1953 foot-and-mouth disease caused a loss, in Europe, estimated at £140,000,000.

[17] Payne, 1962.

In the under-developed countries several other diseases occur, apart from plague, which bring enormous losses to farmers. The development of veterinary science and institutes forms, therefore, a vital link in the work under discussion. In this connection Swedish research workers have made valuable contributions – for instance, in India.

The new chemical means used in the fight against noxious insects will make it possible to extend animal husbandry to regions where it has not until now been possible because of insect plagues. In large parts of Africa, for example, it is the tsetse fly especially which hinders cattle rearing.[18]

SUMMARY

The various measures for improving production are together quite adequate to bring about the increase in production of which Lord Boyd Orr speaks – that 'the full application of science could double and redouble food production'.[19] It is, of course, hardly possible to make an exact estimate of the size of the increase in production, but if we bear in mind that the possibility exists of increasing very considerably the total area at present under cultivation, that the average world yield per acre of the most important foodstuffs is considerably less than half that of western European countries, that the production of foodstuffs in many of the world's warmer climates can go on all the year round, whereas in north-west Europe, for instance, the period of production is limited to about six to eight months, that in many of the over-populated countries it is possible to have three crops a year instead of one, as in most of the north-western European countries, and finally, that there are great possibilities of using the earth's yield more effectively, then it is clear that Lord Boyd Orr's statement is not by any means an exaggeration.

By applying modern scientific and technical methods to agriculture in all those countries where it stands at a low level, it should certainly be possible to increase the production of foodstuffs to such a degree that the food supply would suffice for a world population at least treble that of our present day. More-

[18] Kernaghan, 1963.

[19] Cf. p. 48.

over, it should be possible to achieve this increase in so short a time that the rise in production is greater than that in population. Should the production of foodstuffs need to be increased still more, there are other possibilities. We shall consider these later. Finally, the more animal power can be replaced by motorized implements – for example tractors instead of horses – the more can large areas be freed for the production of foodstuffs for human consumption, and the less devoted to animal foodstuffs.

12

An Increase in Efficiency presupposes Education and Training

THERE are, as we have seen, great possibilities of increasing the production of foodstuffs. But how are these to be realized? Many problems can only be solved by intense research (which calls for very specialized knowledge). We have already discussed this in connection with, for example, ascertaining the state of the soil and its need of nutrients, plant-breeding, the campaign against disease, pests and weeds, and so on. But this is not enough. What is required especially, is that those who cultivate the soil should be properly trained. In short, instructors and consultants are needed. Education, as Alva Myrdal[1] among others has pointed out, plays a very real and important part in the campaign against starvation. The peoples of the under-developed countries must be made literate, for only then will it be possible to instruct them in the techniques required to increase farming efficiency. The need for knowledge among those peoples to whom assistance is being given, so that they might avail themselves of improved techniques has been emphasized in a publication by the Swedish UNESCO Council.

Where necessary, education at an early stage should be directed against prejudices concerning certain foodstuffs, prejudices which are often based on past experiences. Agricultural education should therefore be twofold: it should aim at teaching people to cultivate the soil more effectively, and to use the foodstuffs to the best advantage. In communities with a low but rising standard of living, for example, it occurs that valuable foodstuffs are thrown away because they are regarded as being synonymous with poverty. They have, as it were, been given a mark of class distinction. Such prejudices must, of course, be overcome, as well as the reluctance to use new foodstuffs.

In a category of their own are the prohibitions, based on religious beliefs, against eating certain kinds of food. Many of these prohibitions were originally rooted in concepts of hygiene, and they have been adhered to because they have been incorporated into religious belief.

Questions concerning education were touched upon in various ways during the Rehovoth Conference, and Professor Mogens Pihl[2] of Copenhagen University, suggested that training centres should be established – among other things, for training instructors of different kinds. An important contribution of this kind is now being made by the Scandinavian countries in Tanganyika (Tanzania).

At Kibaha, twenty-four miles west of Dar es Salaam is the site of the £1 million Nordic Tanganyika Project. The project is unique for the Nordic countries as well as for Tanganyika. It is not only the first of its kind undertaken by the Nordic countries anywhere in the world, it is also the first aid scheme undertaken jointly by four countries. The project, which will be completed in four years, incorporates a Farmers' Training Centre, Health Centre, Secondary School and Community Centre.

In the Farmers' Training Centre rational farming will be encouraged by training on model farms. The Centre will accommodate 60 students and, in addition, short 1-3 weeks courses will be given to the local farmers. The soil at the Centre is suitable for cultivating various crops, and the rainfall is comparatively satisfactory. Dar es Salaam is a good market for the disposal of farm products. A nearby forest reserve will make it possible to demonstrate and teach forestry, and a forest nursery will be part of the activities. The nursery will supply seedlings for local forestry development.

Professor Lowdermilk also pointed out in his lecture previously mentioned that a proper use of soil and water resources demands, among other things, not only scientific research and landscape planning, but also the proper training of those who farm the land, and those who are involved in solving agricultural problems mainly concerning the use of soil and water, the

[1] 1961.

[2] 'The Training of Scientific Personnel for Newly Developing Areas'.

cultivation of soil, crop production and animal husbandry. No development programme will be effective, he said, before the farmer himself is efficient. It is only when the farmers can produce more food than they themselves need that people can be freed for other activities in the community. The farmer is therefore the central figure of the development programme.

As things stand it is, of course, not enough to limit training and education to youth. The generation which is already actively engaged must also be taught the elementary knowledge required to apply new methods. Moreover, in order that this level of literacy should be maintained, it is essential, as Alva Myndal has rightly emphasized, that the people in question have access to simple reading matter, for example, a popular newspaper and a farming journal.

13

Food from the Oceans and Lakes

A DISCUSSION concerning the possibilities of increasing food resources cannot ignore the very considerable contribution which can be obtained from the waters of the earth. A rational exploitation of the sea's food resources – fish, sea-weed, etc. – presupposes, however, not only research but also very large capital investment.

Even if fish does constitute an important food item in some countries, it nevertheless plays an insignificant part in the world's total calorie consumption: no more than two or three per cent. According to Yudkin,[1] we are availing ourselves of only 0.01 per cent of the product of the oceans. On the other hand, fish supply an appreciable amount of the good quality protein which is at our disposal. It has been estimated that about one-sixth of the good quality animal protein consumed is fish protein. In Japan it is estimated that fish provide four-fifths of the animal protein consumed.

To help fill the need for protein in many countries the production of fish meal, as made at an experimental factory by the Swedish firm Sunco, could certainly be of considerable importance. This meal, which can be produced at a low cost, is odourless and fatless and contains 85 per cent protein. It can be added to foodstuffs deficient in protein – for example, mixed with wheat flour in bread-baking. All sorts of fish can be used, even so-called refuse-fish, or that used for animal foodstuff. Moreover, the whole of the fish is utilized.[2]

Hitherto, the obvious way of increasing food resources by fish has been to spread information regarding more effective methods of fishing. Norway, to mention one example, is giving such assistance to Kerala, a district in India, by teaching the

[1] 1960.

[2] Cf. p. 55.

fishermen modern techniques and placing up-to-date equipment at their disposal.[3]

However, during recent years several nations – the Soviet Union, China, Japan, and a few others – have begun to fish those great expanses of ocean of the southern hemisphere which have hitherto been more or less neglected. The fishing vessels are supplied with modern refrigeration plants. Along the coasts of Africa, the Soviet Union, together with Poland and East Germany, is conducting extensive fishing of sardines and tunny fish. Along the Brazilian coast Japan, in particular, is carrying out pioneer-fishing on a large scale.[4]

The result of this activity is mirrored in the statistics. For instance, from 1960 to 1961 the fish-catches in the world increased from 38 to 41.2 million tons, that is, by 8 per cent. During 1960 they had increased by 6 per cent. Japan is the leading fishing nation with a catch of 6.7 million tons, followed by Peru with 5.2, China with 5, the Soviet Union with 3.2, the United States with 2.9, and Norway with 1.5 million tons.[5]

At the International Conference on Applied Microbiology held in Stockholm in 1963, Dr M. D. Appleman, of the University of California maintained that it should be possible to increase the yield of fish within the next ten to fifteen years by 60 million tons.[6]

Until now the possibilities of influencing the actual production of fish have been few. However, production can readily be influenced not only in specially prepared fish dams, in which the nutrition can be regulated, but it can also be carried out in lakes where, by chemical means, rotenon, for instance, all fish can be exterminated and the water subsequently stocked with new and more valuable kinds of fish whose nutrition is controlled.

With certain species of fish there is also the possibility of increasing production in the oceans themselves. It is no doubt generally known that the salmon, which is mainly caught in the open sea, spends the first period of its development in spawning-grounds in running water – rivers, streams, and brooks. In many countries attempts are now being made to

[3] Kihlberg, 1962. [4] Borgström, 1962.
[5] Yearbook of Fishery Statistics 1962.
[6] 'Global Impacts of Applied Microbiology'.

improve on the supply of salmon spawn by constructing special breeding grounds. In general, however, this has been done in order to eliminate the unfavourable effects of power plants that have destroyed the natural breeding grounds; but this method could be exploited also with the principal purpose of considerably increasing the stock of salmon in the sea.

It is also conceivable that, in the future, applied science can find methods of increasing the sea's yield of other kinds of fish besides salmon. This was illustrated by an event which took place on the south coasts of Sweden at the beginning of the nineteen-forties. The grass-wrack was entirely wiped out by disease. The result was a very severe reduction in the supply of certain kinds of fish, plaice, for instance, which had been dependent on the grass-wrack for its development. The point to be made here is that it does not appear impossible, in the future, to influence the stock of fish by measures which aim at improving the environment in, for instance, shallow bays and gulfs, which are used by fish as spawning-grounds.

Problems connected with this were touched upon in a lecture by Michael W. Perrin[7] during the Rehovoth Conference. He was of the opinion that supersonic enclosures – by means of which fish can be held within certain limited areas in the sea, for instance, bays and gulfs – should make it possible to conduct fish-cultivation in sea areas (bays, etc.) and not only, as hitherto, in small lakes and fish dams.

Marine vegetation can also play an important role in the provision of foodstuffs. Professor Watanabe from Tokio at the same conference[8] described how, in Japan, marine vegetation is used as food and in other ways in the service of food production. In Japan many different species of algae – a good twenty, brown, red and green – have been used for many years. In the year 1959 the total quantity of dried algae amounted to 340,000 tons, representing a sum of about £20,000,000. To meet the growing need, different methods of cultivating algae have been used for a long time. Some of these algae are used in soups together with beans; others, in a dried form (the so-called *nori*), as a flavouring. Watanabe suggested that algae could play an

[7] 'Food Production in Relation to Present and Future Population Figures'.
[8] 'Utilization of Algae as Foodstuffs and their Potential Use as Green Manure in Japan'.

important part towards satisfying the food requirements of a growing population.

Since the organic substance of algae is relatively difficult to digest, these plants probably have their greatest value as a source of vitamins and mineral salts. Mention should be made, in particular, of their high iodine content, which is probably the reason why strumous ailments occur so seldom in Japan.

It seems very likely that applied science will find a method of improving the digestibility of the organic substance of algae, thereby enhancing their food value.

Of recent years many authorities have drawn attention to the very considerable quantities of organic substance produced in water, and have expressed the opinion that the oceans should be able to make an appreciable contribution to the provision of foodstuffs in the future. It is not only the marine products already mentioned – fish and algae – which can be used to much better advantage, but also the microscopic flora of the waters.

Japanese farmers, and those along the European coasts, have long used seaweed as a form of manure. Of more interest, however, is their practice, in recent times, of using nitrogen-fixing microscopic algae as a means of manuring the rice fields. One species being used for this purpose is *Tolypothrix tenuis*, which is produced in the form of pure cultures, 5 to 10 lb. of which per acre are spread upon the paddy fields. Some experiments have shown 20 per cent increase in yield as a result.

Algae for food can be produced not only in the natural waters (lakes and seas) but also in a more industrial state as will be shown in the next chapter.

14
The Industrial Production of Foodstuffs

AS indicated above, foodstuffs can be produced in other ways than those we have considered up to now, for instance, by means of certain yeast fungi, hydro-carbons and similar organic substances can be transformed into fat and protein. In fact, it is possible to produce organic substances more efficiently through algae and fungi than through higher plants – for instance, by the use of common yeast the total requirement of protein for 140 people can be produced from one acre.[1] Finally, by purely chemical methods, carbo-hydrates, fats and proteins can be synthesized.[2]

Protein Foods From Hydro-Carbons

As mentioned above yeast fungi are very effective protein producers. Through the action of these micro-organisms hydro-carbons, petroleum, are converted into proteins – into a petroleum yeast, as it were.

Experiments with yeast cultivation on crude oil were started in order to find out if the crude oil could be purified from undesirable waxy fractions, paraffin. Then it was found that a substance was obtained, nearly half of which was protein and which contained all the essential amino-acids of the animal protein, and, in addition, several important (valuable) vitamins. As an example it may be mentioned that wheat flour protein contains 1.9 per cent lysin, beef protein 10.0, milk protein 8.7, and the protein in 'protein-vitamin-concentrate' 11.6 per cent. For the important vitamin ribo-flavin the corresponding figures are: 1.0 – 1.5, 2.0, 1-3 and 75.

During the international conference on 'Global Impacts of

[1] Yudkin, 1955.

[2] *Discovery* 8 : 10 and 12, 1947.

Applied Microbiology' in 1963 many questions were discussed which are closely connected with the problem of foodstuff resources; in particular, the possibility of producing more foodstuffs by bio-technical methods. In a lecture to the conference the Director of B.P.'s French laboratory, Dr A. V. Champagnat, who has been in charge of the petroleum-yeast research, said that a medium-sized factory was already manufacturing food from petroleum. There are, Dr Champagnat pointed out, over 700 oil-refineries in the world which already have the basic equipment required for starting the manufacture of protein within a very short time. Expressed in terms of edible protein, albuminous food, it would be possible to obtain 20,000,000 tons per year from crude oil. Only those oil fractions would be used that are less suitable for other purposes.

In a subsequently published paper by Champagnat and his co-workers it was observed that a cow weighing 500 kg. produces, with appropriate feeding, 0.5 kg. of protein during twenty-four hours, while 500 kg. of yeast on crude oil with an admixture of nutrient salts containing nitrogen, phosphorus, potassium and trace elements, in a continuous fermentation process, in a reaction chamber, produces 1,250 kgs. of protein, that is 2,500 times more than the cow. In order to supply the present needs of high-quality protein, 3,000,000 tons per year, 15,000,000 tons of beef, would, according to Champagnat and his co-workers, be required. And by improved animal husbandry it would be very difficult to obtain this result within a reasonable time. But to produce by yeast-fungi 3,000,000 tons of high-quality protein, that is 7,000,000 tons protein-concentrate, per year, only a minor part of the present crude oil production would be required, and it would not take more than a couple of years to build the factories necessary for such production.

With the present world oil production of over 1,300 million tons per year, there is wax (paraffin) enough for production of protein in far bigger quantities than present needs.

There no doubt remains, much to be done before the protein manufactured from crude oil can be turned into a good quality food, but even if the products obtained are not yet very palatable they are, nevertheless, very suitable for mixing with other foodstuffs deficient in protein. It is also very valuable for mixing with fodder for domestic animals.

As the president of the Rockefeller Foundation, Dr J. G. Harrar, pointed out in a lecture at the first plenary meeting of the Stockholm conference, a realm of unlimited possibilities has been opened up through the discovery of the abilities of micro-organisms to synthesize substances which it is difficult to produce in any other way.

The Cultivation of Chlorella

The species of algae which has attracted particular interest in connection with our problem is *Chlorella*, a microscopic, one-celled green algae that is very efficient, for it utilizes solar energy extremely well. Experiments are being carried out in several parts of the world – in India and in Israel, for instance – with *Chlorella* cultures. The organic substance obtained is rich in fat and protein, and the composition can be changed at will by varying that of the solution in which the algae is cultivated and by which it is fed. In other words, a product can be obtained which is either very rich in fat or very rich in protein. The quantities which can be produced are also very large in relation to unit area. According to some statements the quantity of organic substance produced by *Chlorella* is about forty times larger than that which wheat can produce during the same period of time and on an area of the same size. Parker[3] states that, in favourable conditions, the weight of *Chlorella* could increase fourfold in twelve hours and that the annual production per acre could reach 100 tons. From an area of one square mile sufficient protein could be obtained to satisfy about 2¼ million people.

The Production of Synthetic Food

In a discussion on future foodstuff resources, synthetic food substances cannot be ignored.

Already, during the Second World War, synthetic margarine was being manufactured in Germany from carbon and water. The product obtained was certainly qualitatively very inferior to the margarine made from vegetable or animal fat – it had good tenability, but poor taste and lower digestibility, than the natural fat substances – but it was nevertheless edible.

Far more important for the future synthetic production of

[3] 1962.

organic substances, and, consequently, also of foodstuffs, are the results obtained in 1954 from research into the possibilities of producing organic substances from carbon-dioxide and water under the influence of sunlight. In other words, science has succeeded in performing the process of photosynthesis and thereby solved a problem which plant physiologists have worked upon for a long time. By this means man can now use directly solar energy for the production of foodstuffs. From carbon-dioxide and water, in the presence of light and through the agency of chloroplasts (active as a catalyst), starch and sugar are formed.

A statement regarding this achievement was made by Professor Arnon of Berkeley University at a meeting held on December 29, 1954. Plant photosynthesis is that process which makes life possible on the earth, and since it is now possible to perform this process without plants being involved, it is clear that an inexhaustible source of energy and of food is at man's disposal.

The process which is active in photosynthesis is the following: in the presence of chloroplasts the water is broken down by light into hydrogen and oxygen. Hydrogen, with energy received from the sunlight, in combination with carbon-dioxide, forms starch, and oxygen is liberated into the air. Another photosynthetic process and other catalysts have also been found. The latter are, for example, vitamin B, riboflavin, vitamin K, and vitamin C.[4]

Now, as science has succeeded in producing organic substances, starch and sugar, from carbon-dioxide and water, the way lies open for the production of synthetic organic substances – at least, those of the nature of carbohydrates and fats. The consequences this may have for the world's supply of foodstuffs are at present difficult to judge, but it is quite certain that,

[4] According to investigations, photosynthesis proceeds in three stages. First the water is broken down into hydrogen and oxygen (photolysis). Then the hydrogen combines phosphorus and oxygen – which, then, is not liberated – to form adenosintriphosphate, ATP. ATP is an extremely important substance in many biological contexts. It has been described as being the currency with which the cells pay for the deliveries of energy. Finally, ATP is combined with oxygen and carbon-dioxide to form a sugar phosphate, which, in turn, forms starch.

if the processes involved can be reproduced on a technically practical scale, they will prove very significant.

With regard to proteins, mention has already been made of the fact that several of the amino-acids, essential components of protein, have been synthesized.

Many people wrinkle their noses at the mere mention of synthetic food. It should, therefore, be emphasized that when modern scientists speak of synthetic food it is far from being a question of producing food substances in the form of tablets or pills. Those foodstuffs which may possibly be produced in the future from *Chlorella*, from other algae, or from fungi, or purely synthetically, will, no doubt, be essentially similar to those to which we have been accustomed hitherto.

It will undoubtedly take a long time before the results of scientific research mentioned here can be used industrially, nor is there any great hurry. It is, however, encouraging to know that if, in the future, the possibilities of producing foodstuffs in the conventional way are exhausted, there are other possibilities at our disposal.

15

The Supply of Energy

AN essential condition of several of the methods for increasing food supplies which we have discussed is that the energy requirements can be met. Energy is consumed in great quantities in running an irrigation network, in salt water desalination, in the manufacture of fertilizers, and in various kinds of transport. There is, therefore, good reason to consider what sources of energy are at our disposal.

Until quite recently only two sources of energy were utilized on a large scale: fuel and water-power. In comparison with these, wind-power today plays an insignificant role, although it is still used for certain purposes through windmills. A source of energy of more limited, though in some regions very significant importance, is the hot springs, known as geothermal energy. It is utilized on a big scale in Iceland, Italy, California, New Zealand, and in some parts of the Soviet Union. In many other countries research is being done with the same object in view.

Much research has been carried out in an endeavour to harness the energy which finds expression in the ebb and flow of the tides. But until now this work has borne very little fruit, though Soviet engineers are said to be engaged in work on a project to utilize tidal waters for the production of electric power. The planned power-plant is to be an experimental station with a capacity of 1,200 kw situated at a gulf on the Arctic coast north-west of Murmansk. There the difference in water-level between ebb and flow can be more than thirty feet.

Mention should also be made of the experiments to make use of the differences in sea temperature in order to obtain energy.

In many countries water power has been used to a rapidly increasing extent. Even if enormous water power resources are still untapped, the 'stock' of water power is nevertheless limited.

The same may be said of the various kinds of fuel, of which the fossil fuels, coal, oil and gas, are decidedly more important than those which are being continuously created, notably the different varieties of wood which, in general, can be used to better advantage than as fuel. In recent years, in particular, great interest has therefore been attached to the investigations regarding the earth's supply of coal, oil and natural gas.

Before the First World War it was assumed that the supplies of coal, oil and natural gas would suffice for several thousand years. However, this estimate was far too optimistic. Mankind was under an illusion, and it has been necessary to revise radically our ideas on how long the fossil fuels may last. The reasons for this are, among others, that all coal is not worth mining, and that the amount of energy which is required and consumed has risen enormously in recent years; indeed, it has increased like an avalanche.

More than three-quarters of the coal that has been taken from the earth has been mined during this century. Since 1940 more oil has been consumed than during the whole of the history of mankind prior to that year, and the consumption of natural gas has risen even more quickly in recent years. It is this tremendous rise in the consumption of fossil fuels which has stimulated not only the rapid exploitation of water power but also the intensified search for oil.

In this connection it should be mentioned that the average world consumption of energy per person is estimated to be just a little more than the equivalent of 1.4 tons of coal; in the United States the corresponding figure is 8.0, in Great Britain 4.9, and in India 0.14. Even if climatic and other differences between countries affect the requirement of energy to some extent, it is nevertheless obvious that the consumption of energy is closely related to the standard of living.[1] As the latter rises in the technically less developed countries, it is clear that the consumption of energy will also rise enormously.

The earlier optimistic opinion was gradually superseded by a more pessimistic estimate. According to a calculation made a few years ago by the American Energy Commission (AEC), all the hitherto known supplies of fossil fuels would come to an end between the year 2000 and 2025.

[1] BP Information, No. 216, 1962.

However, since that calculation was worked out many new discoveries have been made in many different parts of the world. An American group of research workers is of the opinion that as yet only 30 per cent of the world's total resources of petroleum are known (BP Information, No. 235, 1963). The total supply is estimated to be 200 milliard cubic metres; of which 20 milliard cubic metres have been consumed and a further 40 milliard cubic metres have been located. It remains, therefore, to locate a further 140 milliard cubic metres.

Over and above the oil resources just mentioned there are large quantities to be found in the world's deposits of bituminous shale. Moreover, in the northern regions of North America there are enormous quantities of tar-sand, that is, deposits laid down during the Ice Age containing an oil which is so viscoid that it cannot be worked. However, in the United States it is thought possible to tap this oil by detonating nuclear explosives deep underground. By this means it would be warmed, and become so fluid that it could be pumped up. The same method could be used to tap the oil in the deposits of bituminous shale. In the United States alone it is estimated that 155 milliard cubic metres of oil could be won from shale, and that in the Athabaska region in the province of Alberta, Canada, the tar-sand could yield 95 milliard cubic metres.[2]

The consumption of oil amounts at present to over 1.3 milliard tons per year, and is increasing at the rate of about 5 per cent annually. Even if the increase in consumption were to continue at the same rate, the reserves of oil will be adequate for a long time to come.

In addition to this the known reserves of coal will last, according to Baade, for about another thousand years – assuming that the rate of mining remains the same. In short, it will be a very long time before the supplies of fossil fuels are exhausted.

As long as it was thought that the conventional fuels would be consumed within a not too distant future, atomic energy was regarded as the factor which would prevent mankind from going bankrupt. Today, however, we can safely state that there is no pressing need to use atomic energy before the end of the century. Atomic energy is not, as Baade expressed it, the source of energy for the last decades of our century but could

[2] BP Information, No. 231; 1962, Nos. 233, 234, 235; 1963.

well be the decisive source of energy during the next century.

Once the atom bomb had been constructed it did not take the scientists and technologists very long to work out methods by which atomic energy could be used for peaceful purposes. This led to an intensive search for uranium ores. The reserves now known are so abundant that the prices of new finds have already begun to fall. In Sweden, for instance, there is enough uranium for at least a thousand years.

We could say, then, that to all intents and purposes mankind has inexhaustible sources of energy at its disposal. If, furthermore, methods could be successfully devised by which power could be obtained from hydrogen, then mankind's energy requirements would be met for all time.

The supply of energy was also discussed during the Rehovoth Conference. Three lecturers, Dr Gueron of Euratom[3]; Dr Weinberg of the Oak Ridge Laboratory,[4] and Mr Jukes of the Atomic Energy Authority, U.K.,[5] discussed several problems concerning nuclear power. For example, the possibilities of using low percentage uranium ores, of coupling nuclear power stations to sea-water distillation plants (cf. p. 75), and so on.

At present the capital costs involved in the building of reactor plants are far too high to enable the under-developed countries to acquire them with the purpose of running irrigation plants, desalting sea-water, and so on. But within the foreseeable future conditions will no doubt have changed, and the need for power to solve the problem of water supply may be met in several different ways.

There are, however, ways of developing power other than those already mentioned. One of these is the harnessing of solar energy with which many of the new states are blessed in abundance all the year round. This is the case in the arid and semi-arid regions where the need for power for irrigation is particularly urgent. Although it is only recently that solar energy has been used, the thought of making use of this source of power is by no means a new one. John Ericsson, the great Swedish inventor, constructed the first solar machine nearly a hundred years ago. He recognized very clearly what signific-

[3] 'Atomic and Thermo-Nuclear Energy'.
[4] 'The Possibility of Achieving Energy Autarchy by Use of Breeder Reactors'.
[5] 'Economic Prospects of Nuclear Power in New States'.

ance lay in man's availing himself of the sun's energy. In an essay published in 1868 he asked the question: 'Who can foresee what effect an inexhaustible motive power would have on civilization and on the earth's capacity to fulfil the needs of mankind?' In the same essay he says: – 'In England they already reckon with the time when the supply of coal will give out. Two or three thousand years – mere drops in the ocean of time – will see Europe's coalfields completely exhausted unless mankind takes recourse to the sun.'[6]

At the end of the nineteenth century and during the first decades of the twentieth, little research was done in connection with the utilization of solar energy. But after World War II research was intensified, especially by the Association for Applied Solar Energy in Phoenix, Arizona, and by scientists in Israel.

Sunlight can be transformed into other forms of energy in a variety of ways. At a symposium in Phoenix in November 1955, it was said that atomic power stations with a capacity of 100,000 kw would probably be in general use before solar energy could be harnessed to drive power plants, but that this goal would be attained in the near future. It was also held that smaller solar machines could also be of great importance in sunny, warm countries – for instance, to heat living accommodation and household water, to run refrigerators, air-conditioners, irrigation pumps, and so on. This would mean more and better food for more than half of the world's population, would increase the facilities for storing and preserving foodstuffs, and would improve health and the capacity for work.

At the time of this symposium solar energy was still far too expensive, but the opinion was expressed that continued research would find possibilities of making it cheaper. This hopeful attitude did not have to wait long to be proved right. Until then the solar machines in use had been uneconomical, in particular because they could only make use of a very small part of the solar radiation. Far too much was lost. A short time after the meeting in Phoenix, scientists in Israel succeeded in constructing a solar machine which, relatively economically, could be used for the production of steam. The machine consists of a number of specially prepared plates heated by the sun

[6] Goldkuhl, 1961.

and giving off heat to water moving in coils round the radiating plates. This method makes use of 16 per cent of the total of the solar energy accepted. A few years later the method was developed further on an industrial scale in a plant in the Negev. Steam is now being produced there at least as cheaply as by conventional fuels.

Another Israeli apparatus to make use of solar energy is based on glass made so that it accepts warmth through one side but reflects radiant heat from the other: a sort of warmth accumulator, as it were. It is said that houses might be heated in this manner.

Devices for heating water by solar heat are already being manufactured industrially in the United States, both for the home market and for export. A similar manufacture is being planned in many other countries. In Israel solar energy has already been used in practical life on quite a large scale. In the new town of Eilat on the Red Sea every house is equipped with a small solar machine to drive air-conditioners, refrigerators, and so on. There are similar constructions in the United States, Japan and South Africa. A survey of the practical uses to which solar energy has been put is given by Ohman and Goldkuhl.

A new construction, devised by Dr Tabor of Jerusalem[7] – the 'solar pond' – aroused great interest at the Rehovoth Conference. This is a basin with a black, flat bottom, filled to a depth of about three feet with water. At the bottom the temperature of the water can rise to 90°C., whilst the temperature of the water at the surface is only slightly raised. The calorie energy in the bottom layer can be transformed into other forms of energy in a variety of ways. Dr Tabor had calculated that such a solar pond, with an area of one square kilometre, could generate 6,000 kw of power. Making use of present techniques for transforming energy it has been estimated that a plant of this size would cost about £400,000.

Soviet scientists are said to be working on a project for a large solar energy plant in Armenia which would produce 20,000 tons of steam for a turbo-generator with a capacity of 1,200 kw.

Solar energy has the advantage over atomic energy in that it requires less expensive installations, and above all does not

[7] 'Solar Energy Research and the New States'.

involve any of the radiation hazards of the kind associated with atomic energy.

From the above it is clear that the necessary supply of energy may be regarded as being guaranteed for all time. It is now only incumbent upon us to make use of this energy to the best advantage, and to use the most suitable source of energy in each specific case.

16

The Need for Aid. Different Forms of Assistance

The Distribution and Donation of Foodstuffs

The account given hitherto shows quite clearly that there is every possibility of being able to produce enough food for a world population several times greater than at present. It also makes clear that time is needed before the necessary measures can be applied in all those regions where there is need, before they can give the desired results. Consequently, the fact that, at least for the immediate future, the peoples of many countries will still be suffering from hunger and starvation, whilst in other countries there are surpluses of unsaleable foodstuffs, must be taken into account. These should, of course, be used to alleviate the suffering of those in need.

In the past it used to happen that an uneven distribution of food would be 'corrected' by migrations; that is, people moved away from the districts short of foodstuffs. The last sizeable migration caused by acute shortage, was, no doubt, that which took place in the middle of the nineteenth century from Ireland after the severe famine which resulted from the complete failure of the potato crop in 1845 and 1846 – at that time the Irishman's most important food.

Any migration which takes place today – from over- to under-populated countries – is quite insignificant. Instead, quite considerable transportation of food takes place. I do not refer to regular international trade, but to the gifts of foodstuffs to starving people from countries with surpluses. In such cases it sometimes happens that dietary habits prove a hindrance. Here is an example. Rice is the most important foodstuff for more than half the population of the earth, but in many rice-growing countries there is a shortage of food. Moreover, there is no significant surplus of rice in any area, though there are surpluses of wheat in many countries. It has therefore been necessary in some cases to accustom starving people to a diet of wheat instead of rice.

During the last two decades countries with surpluses of foodstuffs have put them at the disposal of countries that have needed them but have not been able to pay for them. The United States began by sending wheat to India in order to prevent famine in a region where the crop had failed. Since then, aid of this kind has grown considerably. Under the famous Public Law 480, 1954, the United States has donated foodstuffs to a value of 11 milliard dollars to forty different countries, or sold them on terms advantageous to the buyers. For instance, the payment has been deposited in the land receiving the goods to be used for investment purposes.

The purpose of these donations has varied; it may, for instance, have been to prevent famine, or to hasten economic and cultural development by enabling the recipient to liberate labour for building roads, schools, dams, and so on. This activity now forms part of the 'Food for Peace' programme which was begun in 1960, according to which 1,200 agricultural specialists have been stationed in fifty-nine countries.

During recent years UNICEF has started a food programme in several African states by distributing dried skim milk among the schools. WHO has also made considerable contributions to the furtherance of a better diet in African countries.

It may seem a simple matter to send foodstuffs from one country to another, but, in point of fact, not only are there certain economic problems, but there is also the difficulty of distributing the food to those who need it most. In reality, very careful organization and good storage facilities are both essential.

An uncontrolled import of food can occasionally cause difficulties in the receiving country. The surpluses must, therefore, be chosen and distributed in such a way that the aid given does not interfere with normal trade. From this point of view a very suitable foodstuff is powdered milk, which is used in some poorer countries for school meals.

With the purpose of preventing the increasing bilateral aid (Government to Government) from interfering with world trade in foodstuffs, FAO worked out internationally recognized directives. Later in 1962 FAO/UN developed the 'World Food Programme' which is to work with this aim in view. A part of this programme is the 'Development through Food' plan which aims at enabling projects to start which can further production.

Even if the donation and distribution of foodstuffs – which the last mentioned programme has made much easier – can bring very valuable help, they do not constitute a final solution of the problem of food supplies. The proper long term solution is to help the under-developed countries to produce more food themselves.

Or is it necessarily *food*? This is a vital question, for sometimes another question is asked: Is a hungry people in the tropics to cultivate and export cotton, rubber, cacao, sugar and jute in order to be able to buy wheat, meat, dried fish and other foodstuffs, or is it to cultivate cereals, beans, vegetables, etc. for its own sustenance? The usual answer is that the former procedure should be followed. But, says Pirie, this answer is not always convincing – and for two reasons.

Several of those countries in which people are starving were first developed as colonies by people who were interested in such industrial raw materials as those just listed, but had little interest in the native population's need of foodstuffs. However, much has changed since those days.

The techniques involved in cultivation are for the most part more developed in respect of crops for industrial purposes than for foodstuffs. Occasionally the opinion is voiced that it would be better to discontinue the production of industrial crops in many parts of the world, and to intensify the cultivation of food crops – and, moreover, to cultivate very largely those foodstuffs that are rich in proteins and vitamins rather than those rich in starch and fats.

This concept may well be justifiable with regard to many countries, but it should not be forgotten that, in tropical countries in particular, there are extremely favourable conditions for the cultivation of industrial crops, many of which are obtained from perennial plants, trees or shrubs, whereas conditions are less favourable for the cultivation of annual food plants. In such cases it is obviously justifiable to concentrate on the cultivation of crops for export, and to meet the food requirements by importing foodstuffs. Hawaii is a good example of the value of such specialized production: there the two main crops are sugar-cane and pineapples.

In this connection it should be emphasized, however, that it is essential that the advanced industrial countries should frame their trade policy in such a way that the countries producing the raw materials are favourably treated.

Education, Enlightenment and Advice

An important part of all projects to increase the production of foodstuffs, is, as already mentioned, the provision of education, enlightenment, advice and guidance.

At an early stage of its activity FAO included various forms of technical assistance: experts and technically trained personnel were sent to under-developed countries to teach the farmers the use of artificial fertilizers, better seed and modern methods of cultivation.

During the twenty years which have passed, since FAO began its activity, more than 15,000 experts have been sent out to countries in need of aid by the whole United Nations Organization, of which some 3,500 were sent from FAO to sixty countries. At present FAO has about 1,000 experts working in different countries.

In the period 1950 to 1963 12,500 experts were sent out by the United Nations Technical Assistance Programme. Together these 'assistants' have served 28,877 'expert-years', of which FAO's share is 7,978. The United Nations Organization has also granted about 30,000 scholarships to people in under-developed countries, (over 4,000 of which were given by FAO) for training in Europe or America to enable them to lead various projects in their homelands. Another form of assistance is the organization of collective studies.

No less than 165 countries have benefited from this kind of help. Furthermore, FAO has helped some governments to introduce the study of nutrition into the school curriculum and also to organize school meals. In addition, schoolchildren have been taught gardening and chicken rearing. There is every reason to believe that measures like this will bring about an appreciable improvement in diet.

Capital Requirements and the Financing of Assistance

The advice and enlightenment given by experts, the organized courses and so on, have certainly been invaluable, but not always completely satisfactory. For example, in arid and semi-arid regions the results have not come up to expectation, notably because there were no possibilities of meeting the need of an adequate water supply. For this considerable capital investment is necessary. Furthermore there were no fertilizers nor improved seed varieties available.

The peoples of the under-developed countries do not need advice and guidance alone, they need also the ready opportunity to follow the advice given. It is not only the knowledge of how food should be produced which is lacking in the under-developed countries (in several, India for instance, the requisite knowledge is already there), but in all these countries there is a lack of capital. To achieve the necessary increase in production which, above all else, involves the development of water and irrigation systems, and also the building of power-stations, factories for the manufacture of artificial fertilizers, and so on, enormous capital-investment is required. So that such investments should materialize in the technically undeveloped countries, co-operation between the technically and scientifically advanced lands is essential. Separate contributions from small nations can only be like drops in the ocean. An international organization is needed through which large capital sums can be placed at the disposal of the countries in need – not, in the first instance, in the form of gifts, but as long-term loans. Such assistance towards self-help can enable the needy peoples to improve their food supplies themselves, and also to solve their social and cultural problems. Admittedly, to start with this would have to be done with outside help – and this assistance must be given free of charge.

If we take this as our point of departure and look at the measures which have been taken hitherto, we shall reach the conclusion that they are far from adequate.

As de Castro, among others, has pointed out, it was already clear when FAO started that if the organization was to fulfil its task it would need greater authority than that for which provision had been made in its statutes. The endeavours made by Lord Boyd Orr, the first Director General of FAO, to obtain the necessary authority, a food supply council for the world, were, unfortunately, in vain. FAO has certainly done very meritorious work, but this has not been enough to solve the world supply problem. Indeed, when de Castro resigned in 1955 as chairman of FAO's council, he said that he felt the work they had done had been disappointing and that this was, among other things, attributable to the fact that they had not been able to devise a realistic foodstuffs policy answering the needs in question.

To provide assistance to the under-developed countries, UN and its various specialized organizations have two programmes in common, namely EPTA and SF. During 1962 EPTA had about £11,000,000 at its disposal. By July 1963 SF had granted £100,000,000 towards 327 projects in 120 countries since its inception in 1959. The countries in question contributed £133,000,000 towards the costs involved.

Sweden's contribution in 1962 was in the region of £540,000 to EPTA and £2,000,000 to SF. Apart from this, Sweden contributed to the assistance programmes of the various UN organizations by means of membership fees. In the same year, Sweden sent £180,000 to UNICEF and £238,000 to World Food Programme.

In relation to the tremendous need for assistance, not least the need for capital, the aid given by UN is far from enough. However, large sums of money are also placed at the disposal of the poorer countries in the form of loans for major development schemes. Capital investments for a variety of projects are made by the International Bank for Reconstruction and Development – the 'World Bank' – which was established in 1946, and which, since 1948, concerns itself mainly with the under-developed countries. Up to 1963 the cumulative total of loans advanced by the International Bank amounted to about £2,410,000,000, of which two-thirds have gone to the under-developed countries. But even this is not sufficient.

Great Britain and several of the Commonwealth countries are giving assistance within the framework of the Colombo Plan for south and south-east Asia. This plan is based on collaboration between the states situated within the region in co-operation with other Commonwealth countries. It has been described as a 'symbol' for aspirations and expectations. The Colombo Plan assistance programme may be described as 'bilateral'. During recent years private investments have amounted to about £100,000,000 a year, and Member-Government contributions to a figure somewhat higher. The total sum is about £240,000,000, which corresponds to 1¼ per cent of the national income of the contributing countries. Between the end of the war and 1960 (inclusive) the Colombo Plan has rendered financial assistance amounting to a cumulative total of £2,500,000,000. Notwithstanding these efforts such assistance is still inadequate.

The largest assistance programme to have been set up so far is the Point Four Programme of the United States. In 1949 President Truman made this American investment and assistance plan public. Its aim, like that of all the United States' assistance activities, was formulated in an Act accepted by Congress in 1950. The purpose was stated to be:

(1) To develop the productivity and purchasing-power of the countries concerned;
(2) To create new sources of income;
(3) To raise the standard of living.

The Point Four Programme has been described as a systematic attack on 'the vicious circle'. The underlying idea is that progress must begin in agriculture – that is, sufficient quantities of food must be procured – and that all else will follow, in many cases of its own accord. Development in the United States has been held up as an example. About the middle of the nineteenth century 85 per cent of the population were engaged in agricultural work to provide food for the nation, only 15 per cent were free to devote themselves to cultural and technical development. Today conditions are reversed; less than one-third of the population needs to be concerned in the production of food.

The administration of the Point Four Programme was entrusted to a special branch within the American Department for Foreign Affairs, the Technical Co-operation Administration (TCA). In 1951 this programme was combined with the Mutual Security Act (MSA).

During the years 1950 to 1953 the Point Four Programme rendered assistance to thirty-five countries in southern Asia, the

Near East, Africa and Latin-America, to the value $277,000,000.

That part of the programme concerned with giving help to the Asian countries came particularly into focus early in 1955 through Mr Stassen's journey to that group of countries which in America are called 'the free arc of Asian countries', that is, the countries from Pakistan in the west to Japan in the east. On his return Mr Stassen, former Director General of the United Nations Foreign Operation Administration (FOA), proposed a programme for increasing assistance to those countries. Soon afterwards the United States' Congress, at the suggestion of President Eisenhower, raised the allotted sum by $200,000,000 to a total of $915,000,000 for technical and financial assistance to those countries. In 1955 a total of 766,000,000 people lived in these countries, nearly a third of the world population at that time. The American assistance programme had been formulated as an eight-year-plan, and the total sum just mentioned was allotted to the year 1957. The purpose of this assistance was to raise productivity in the benefiting countries by about 30 per cent. Included in the programme was the provision of raw material for the production of atomic energy for peaceful purposes.

In a leading article of the *New York Times* it was said that enormous sums of money were certainly involved, but that such assistance would in any event be far less costly than even a limited war.

This money is being used, in the first place, for basic economic development, that is, power plants, irrigation systems, improvements in transport on both land and sea – as in the case of the Damodar Valley – and factories for the manufacture of artificial fertilizers. In other words, the primary aim of this Assistance Programme is to increase the production of foodstuffs and to create a stable economic foundation for social and cultural advancement.

In my opinion it is particularly valuable that such stress has been laid upon the fact that it is, above all, the production of foodstuffs and other essential commodities which should be furthered. For there is often talk of the need for improving and developing the industrial capacity in the less advanced countries, whereas anyone who has occupied himself with problems of the under-developed countries knows that it is first of all the foodstuff resources which must be increased. The problem of food supplies can never, as already pointed out, be solved by the transport of foodstuff from countries with surpluses. The reader is reminded here of Dr Raman's statement cited earlier (see page 31).

'A program such as this,' says the *New York Times* (January 19, 1955), 'should do something, at the outset, to disarm the critical charge that the United States is interested only in military assistance to Asian states. None of this program falls into the strictly military area, although it goes without saying that economic strength is the backbone of military strength.'

The Four Point Programme aims first of all at raising the standard of living of these Asian countries and, by so doing, making freedom more valuable and more worth defending. Much of that which is envisaged can, in the long run, be achieved by means of private investments. But if the programme is to be carried out at the pace demanded by the great and immediate needs, it is apparent that the United States government's intervention was necessary.

'It is to be hoped, nevertheless,' says the *New York Times* in the leading article already referred to above, 'that what is done in this fashion will stimulate the flow of risk capital and change, in due course, the entire picture. This, in turn, will depend on attitudes and climates of opinion in the countries that need help.'

In recent years the assistance given by the United States has increased considerably. Moreover, whereas at an early stage it consisted largely of donations, now loans constitute the major portion. However, the contributions towards multilateral aid still play a relatively minor role. During 1959, for instance, bilateral donations and loans amounted to $2,270,000,000 whereas the contributions towards multilateral assistance amounted to only $60,000,000.

American assistance to the under-developed countries has, in recent years, been carried out in the main within the framework of three important programmes. One of these, the most important, is based on the Mutual Security Act (MSA) and is, in part, of a political and military nature. The second is the already mentioned food assistance programme in accordance with Public

Law 480, 1954. The third is in the form of an expansion of loan grants through the Export-Import Bank.

Gifts and loans in accordance with the MSA programme were granted earlier through the International Co-operation Administration, ICA, and the Development Loan Fund, DLF. In the autumn of 1961, these were amalgamated into the Agency for International Development, AID. Established in 1957, DLF soon became the most important loan fund for development projects.

MSA assistance undertakings amounted to 2.7 milliard dollars during the period July 1, 1957, to June 30, 1959, and $850,000,000 were in the form of loans. No assistance of a purely military nature is included in these figures. During the same period the undertakings in accordance with P.L. 480 amounted to $1,118,000,000, and loans from the Export-Import Bank amounted to $1,019,000,000. The total amount from all three programmes during these two years was, therefore, just over 4.8 milliard dollars.

The assistance rendered by the United States is now mainly handled by AID which, in 1962, gave economic support amounting to 2.5 milliard dollars. In the same year 1.6 milliard dollars was at the disposal of countries in need of assistance in accordance with P.L. 480.

The United States government supports the separate organizations concerned by granting assistance in different ways – amongst others, in the form of export credit guarantees.

From Canada and several European countries – France, West Germany, the Netherlands, Switzerland, the Scandinavian countries, the Soviet Union, and a few other East European countries – large sums of money are given both as donations and as loans for bilateral assistance in many different countries. The value of the bilateral assistance which advanced industrialized countries are giving amounts to a total of about 2.75 milliard pounds sterling a year. Approximately one fifth of this consists of private investments (UNESCO *Courier*, March, 1963).

In the leading article of the *New York Times* to which reference has already been made,[1] it was stated that the results of the assistance envisaged would depend on the attitude of the peoples in need. Here we are really faced with the crux of the matter. The desire to co-operate, and an honest striving to make

[1] Page 126.

use of the assistance offered to the best advantage are, of course, absolutely essential if the undertaking is to be successful. In consideration of this it would, in my opinion, be better if the assistance offered could go via UN to a greater extent than hitherto – that is, via that international organization of which the countries who are receiving assistance are also members. This opinion is based on a number of observations. For instance, in many parts of the world, such powerful countries as Great Britain and the United States, are still looked upon with a certain distrust. It is true that it is not political imperialism which is feared, but, although unjustified, there is a fear of economic imperialism. There is little doubt that people who are representatives of the smaller countries can reckon with greater goodwill among the peoples in need of assistance than can representatives of the great powers. To channel the major part of such assistance through UN and its various organizations would eliminate the practice of grouping countries into those who render assistance on the one side, and those who are in need of assistance on the other; it would be more a matter of 'World mutual aid'. This suggested change in terminology is based not only on psychological motives but also on fact. There is no doubt that if assistance is not given and received quickly enough, then difficulties can well arise also in those countries which are at present relatively well established and which are not counted among those in need of assistance.

It is therefore of very great importance to build up the international organization, under the auspices of UN, which is equipped with such capital resources that projects to further production can be started in many under-developed countries.

Such an organization was suggested in UN in November 1951, by Leo Mates, of Yugoslavia. In the following year UN's general assembly commissioned the Economic and Social Council to work out a detailed plan for the establishment, as soon as possible, of a Special Fund for low-interest, long-term loans to under-developed countries with the purpose of assisting them, to the extent they may wish, in hastening their economic advancement; also to finance projects of basic importance for such development.

A committee of experts set up for this purpose submitted a formal report in March 1953 in which the establishment of a

Special United Nations Fund for Economic Development, SUNFED, was unanimously recommended. It was suggested that this Fund should begin functioning when contributions from the various countries covered the needs of a two-year period, and that thereafter contributions should be made according to the results obtained, the needs existing, and the amounts governments found it possible to contribute. All contributions should be voluntary, and each country was to decide how much it could afford. Any idea of a compulsory charge according to ability to pay was rejected.

As it was considered essential that the fund should have sufficient capital at its disposal to make a good start, the committee recommended that contributions should be underwritten to an amount of at least $250,000,000 by at least thirty countries in order to cover the need during the first two-year period.

However, this interesting proposal was met with criticism. For instance, it was pointed out that a pilot project of *only* $250,000,000 was not an adequate enough reason for establishing a special international organization of the type envisaged. Further, that there already existed 'interlocking' assistance programmes – the Expanded Programme of Technical Assistance, the Colombo Plan, the Point Four Programme of the United States – and that the creation of yet another fund was unnecessary. To these arguments it was answered that SUNFED could be the unifying organization within which the other funds and programmes could co-operate.

The greatest obstruction, however, was the unwillingness of the United States to participate until no definite progress had been made in the field of world disarmament. Many other countries assumed the same attitude. However, since $250,000,000 is but a small fraction of the United States present Defence Budget, there should not have been any reason to wait for disarmament before freeing money for SUNFED. It may be added that the sum which would have made up the contribution from the United States would have been very insignificant in comparison with their total Foreign Aid Budget – which then amounted to 3.5 milliard dollars – in fact, even compared with the assistance they were giving the Asian countries.

Recognition of the need for large capital investments of the kind for which SUNFED was intended increased, however, and an expression of this is the establishment of the International Development Association, IDA. At the beginning of 1960 the statutes of this fund were presented to the governments of those countries which could become members, that is, the states who were members of the World Bank and of the International Finance Corporation. According to these statutes IDA was to begin to function when $650,000,000 of the proposed initial capital of $1,000,000,000 had been subscribed.

The task of IDA is to further the development, increase the production, and thereby to raise the standard of living in those less developed regions of the world where the countries are members of the organization.

Whereas, on the one hand, the World Bank mainly grants loans to governments or against a State guarantee, and the International Finance Corporation mainly invests in private enterprises without State guarantee, IDA, on the other hand, grants loans at low interest, without State guarantee, for the financing of projects of high development-value, even if these cannot immediately bring in an income. In other words, the function of the IDA is the same as that which SUNFED should have had. UN's Special Fund (SF) has similar aims. This organization is financed by voluntary contributions from the member States. It grants loans for projects which precede capital investments and which are of fundamental importance for advancement.

During recent years UN's activities have progressed in an encouraging way in several respects. For example, the plan proposed for the economic development of Africa by UN's Economic Commission for Africa (ECA) at the Commission's annual meeting held in Tangiers in the beginning of 1960. The Commission decided, among other things, that the programme for 1960 to 1961 should concentrate its attention on measures aimed at furthering the development of:

(1) Agriculture and Handicraft,
(2) Industrialization and Transport,
(3) Stabilization of the prices of raw materials,
(4) Financing of Industry and Commerce,
(5) Statistics and Research.

The opinion was also expressed that a point in the programme to which particular emphasis should be given was the training and education of Africans so that they themselves might assume their rightful role in the work of developing their own continent. UN's various organizations and all those countries which are members of UN are urged to further this development in every possible way.

The Commission also approved a scheme to map Africa's natural resources. Such a scheme is of inestimable value, not only for the industrial development of the continent but also for the development of agriculture in many regions.

The question of the financing of assistance to the underdeveloped countries was discussed in detail during the Rehovoth Conference. The Director-General of the Bank of Israel, David Horowitz,[2] analysed the fundamental conditions which have to be fulfilled in order that the new states may develop fruitfully. Among other things, he emphasized that an economic breakthrough – achieved by importing capital and knowledge – is, under present day conditions, the only alternative to the downfall and eradication of a democratic way of life. Investments in the less developed countries by the more highly advanced nations have been far too small up to the present – with the exception, it is true, of those made in countries rich in oil resources. Extensive assistance would be beneficial to under-developed and advanced countries alike, and it would be possible without undue sacrifice on the part of the latter. Even if only a small fraction of the national wealth of the advanced countries were to be invested in those at present under-developed, it would soon lead them to self-sustained economic growth.

Professor Carl Iversen of Copenhagen[3] pointed out that it is a great mistake to believe that economic development is synonomous with industrialization. In many countries a modest beginning in the improvement of agriculture can give more valuable, though less spectacular results. Moreover, and this is an important point to bear in mind, far smaller investments are required for this purpose, at least in the early stages, than for the development of industry.

[2] 'Politics and Economics in the New States'.
[3] 'Problems of Inflation in Developing Countries'.

An important aspect, which must be considered, and which was touched on by many speakers at the Conference, is the form in which financial assistance should be given – as loans or as contributions. Many countries unquestionably prefer to receive assistance in the form of loans. However, a representative for Nigeria pointed out that in some countries there is a certain reluctance to accept the conditions which are attached to loans. He mentioned also that his country possesses tremendous potential wealth in the form of fertile soil, coal, oil, and so on, and that it had been estimated that, in order to utilize these resources, the necessary capital investment would amount to £150,000,000. So far, the financial support gained up to then, partly from the World Bank, amounted to scarcely one-sixth of that sum. Other speakers were of the opinion that the situation in many countries is often so serious that it is quite inevitable that, at least to begin with, any assistance rendered would have to be in the form of donations.

In a lecture delivered to the participants of the Conference by the Directors of UN's Special Fund, Paul Hoffman, it was pointed out that enormous sums of money had been wrongly used, because assistance had been given with the underlying motive of gaining political influence in those countries to whom money had been given or lent. Hoffman was of the opinion that financial assistance should certainly further a political goal, but that the countries giving assistance should set a high goal – that is, to enable the under-privileged peoples to choose voluntarily the economic and social system under which they wish to live. The best method of counteracting irrelevant influence is to give assistance as far as possible through the international organizations, and not in the form of bilateral projects. One very important reason for this is that the countries receiving assistance would then feel themselves to be an integral part of the campaign and would, therefore, feel encouraged to assume a role of greater responsibility.

This suggestion is especially applicable to the larger financial assistance plans, and should not, of course, prevent individual countries or enterprises from shouldering the responsibility for more limited projects, such as the establishment of schools, hospitals, model farms, and so on. Nor should it prevent individual firms and organizations from investing capital in the under-developed countries with the purpose of developing this or that branch of industry.

One of the most urgent needs at present – and this is particularly valid with regard to measures in the sphere of agriculture – is a very appreciable increase in financial contributions to enable measures to be taken of a pre-investment character that is, investigations which must be undertaken in order to prepare the ground for sound investments. All too often capital has been squandered on projects which have not been thoroughly investigated and planned, so that the results have fallen short of what had been envisaged.

In an excellent summary of the planning from a scientific and economic point of view, Professor Bergmann of Jerusalem 'Planning of development and Research in the New Countries' emphasized that the new countries should try to imitate the history of human development as a whole, only they must develop more rapidly. To enable them to achieve this, the need which must be met first and foremost is that of a well balanced diet of good quality. As has been mentioned already, the food in the majority of the new countries – and also in many older ones – is not only insufficient but also deficient in protein, and so does not permit a high level of human productivity. There are, to sum up, three needs of prime importance:

(1) Expansion and regulation of agricultural production;
(2) Survey of the natural resources; both plants and minerals; and
(3) Industrial development, first of all in utilizing agricultural produce and agricultural surpluses.

The progressive stages of development in the various branches of agricultural, commercial and industrial life must, as has already been emphasized in another connection, be so planned that they run parallel. For instance, as man-power is freed from agriculture it must be possible to provide work in some other activity so that unemployment is avoided.

17

Food Supply, Disarmament and World Peace

THE amount of money which can be invested in the less developed countries is, of course, largely dependent on disarmament. There is, in this respect, a clear connection between assistance programmes on the one hand, and plans for disarmament on the other, to which the United States has drawn attention (see page 129). Should a universal and comprehensive decision on disarmament be reached, it would obviously increase the possibilities of investing capital for productive projects. It is true that the Soviet Union and the Democracies of the Western World are able to render appreciable assistance even as matters stand at present, but the assistance which is now being given is a small fraction of what could be given after complete world-wide disarmament had been achieved. According to a UN report entitled 'Economic and Social Consequences of 'Disarmament' (1961), the total world expenditure on armaments amounts to about 40.5 milliard pounds. Even a much smaller amount than this would, if used for productive purposes, rapidly bring about radical changes.

Disarmament has in many quarters long been cherished as an urgent goal, but nevertheless still a very distant one, more a matter of wishful thinking than of practical politics. But there are many signs which indicate that the latest developments in weapon-techniques are calling upon mankind to come to its senses. The hydrogen bomb is not the worst menace. In a leading article of March 13, 1955, under the headline 'The Earth Bomb' the *New York Times* wrote: 'As . . . we struggle to formulate a policy which will protect at once freedom and peace, it becomes clearer that what we have to fear – what mankind has to fear – is not the puny, old-fashioned A-bomb, nor its successor the H-bomb, nor the H-bomb encased in cobalt; it is rather the explosion of the earth's civilizations in a final catastrophic war; it is the Tellus bomb as we might call it.'

'East or West, wherever we turn, we must be firm and clear in our concepts and statements of what we will defend and on what rivers or shores or islands we mean to take our stand. We must be equally firm in our announced desire and drive for peace. We must be infinitely patient. We cannot talk like truculent little boys in a school-yard during recess. The burden of the future rests upon us. Neither fear nor hate can guide us. We must stand with as little passion as possible for the twin objectives of peace and freedom.

'This is a supreme moment in the history of humanity. The earth is tinder. Our policy must be to see to it that nobody puts a match to it.'

Harold Macmillan, then Minister for Defence, spoke in the same vein on March 2, 1955, in answer to a statement made by the then Leader of the Liberal Party, Mr Clement Davies, who was also the Chairman of World Association of Parliamentarians for World Government:

'On the whole question of disarmament, our purpose is simple and our record is clear. Genuine disarmament must be based on two simple but vital principles. It must be comprehensive and it must provide a proper system of control. It must be comprehensive, by which I mean that it must include all weapons, new and old, conventional and unconventional. The control must provide effective international, or if we like supra-national, authority invested with real power. Right Honourable Members may say that this is elevating the United Nations, or whatever may be the authority, into something like world government. Be it so, it is none the worse for that. In the long run this is the only way out for mankind.

'Of course, it will take time and patience to reach our goal. We have had disappointments, we have had setbacks, but we shall persevere. For disarmament is two things – and when we achieve it, it will be this double triumph: it will be both the symbol of peace and the means of preserving it.'

Similar statements have since been made by many prominent statesmen, for example, Earl Attlee, John F. Kennedy, Dag Hammarskjold and U Thant. In a speech given in 1962, Mr Gaitskell outlined a detailed plan for the transformation of UN into a World Authority which, with the support of a worldwide law and permanent police force, would be able to uphold peace in a completely disarmed world.

It is extremely valuable and encouraging that so many leading statesmen have so clearly formulated what so many people are longing for!

To Macmillan's concluding statement could be added the fact that disarmament would lead to a higher standard of living for all the peoples of the world, since it would obviously enable us to increase the production of foodstuffs very considerably and also to increase our power resources for peaceful purposes.

We could say, in fact, that the problems involved in the world supply of foodstuffs, of worldwide disarmament, and of world peace are, even if not completely interdependent, nevertheless very closely bound up with one another. It was, therefore, a valuable move when, on March 19, 1955, President Eisenhower named a special Secretary for Peace. It was Mr Stassen who was named to fill this post when, on July 1, 1955, FOA, the organization for which he had been director-general, ceased to function and handed over its activities to the State Department.

This probably was the first time in the history of mankind that a government appointed a Secretary for Peace and the day should, as the *New York Times* suggested, be regarded as an important historical date. In an article entitled 'Secretary for Peace' this paper wrote: 'It is not naïve to appoint a "Secretary for Peace". The noble dream of past generations has become a necessity for survival. The prayers of the nation will be behind Mr Stassen and the hope he stands for.'

It is most profoundly to be hoped that all negotiations being made to bring about disarmament within the near future may arrive at a successful conclusion. Once such a goal has been reached it will be possible for mankind to create a 'paradise on earth'.

18

Conclusion

THE extremely urgent task confronting our generation – to produce food in sufficient quantity and of adequate quality for the earth's rapidly increasing population – may appear terrifyingly great and very difficult to solve. But a close survey of the possibilities of increasing food supplies makes quite clear that enough food *can* be produced for a population several times larger than at present. The question which was raised at the beginning of this book, 'Can the Earth feed us?', may therefore be answered in the affirmative. The necessary condition is, however, that mankind allows common sense to rule and creates the foundation for a peaceful co-existence.

There are practically no limits to the possibilities of producing food on this earth, and even so, we have a long way to go before we reach them. In reality, we are at present utilizing only a small percentage of the earth's capacity. The production per area unit is in some of the intensively farmed countries (see Table IV) several times greater than that of most of the large agricultural countries and of practically all the under-developed countries.

By a worldwide adoption of methods which are already being used in some countries on a large scale, and by making use of new scientific procedures in the production of foodstuffs, much larger quantities can be produced than would have been envisaged only a few decades ago.

With very few exceptions, the poorer countries where starvation is prevalent have the basic conditions necessary for the production of more food: fertile soil, water and labour. Knowledge and guidance in agricultural matters are already available to most of these countries, and where it is not, it should be made available in increasing measure. Other measures must also be taken. Dams and irrigation systems, roads and railways must be built; factories for the manufacture of fertilizers must be built and equipped; plant breeding, plant protection and animal husbandry, including veterinary work, must be organized; schools and other training and research institutions must be established to educate and train a new generation of farmers; hygiene and medical care must be developed. Considerable capital is required for many of these essential measures, and this must be placed at the disposal of the poorer countries either in the form of loans or gifts. Without capital it will be possible neither to begin nor to hasten the development towards higher productivity. With such assistance, the needy countries could reach a higher standard of living within a reasonable space of time.

The world foodstuff problems should, in the first place, be solved in such a way that the food is produced where it is needed – provided, of course, that the natural conditions are favourable; and, in the second place, by such a development of international trade that those countries which have good conditions for, let us say, the production of industrial raw materials are given the facilities to meet their food requirements through imports.

Large investments are also needed in order to utilize the enormous supply of foodstuffs in the world's seas and oceans, and also for developing the industrial manufacture of food.

The view that the problem of feeding the world population can be solved only by birth control is utterly wrong. The essential thing is that, by international co-operation, the earth's actual and potential capacity for production is fully utilized. For the real reason why so many countries are under-developed, and why so many people are starving, is that the earth's resources are not used to the full. The problems involved in the supply of foodstuffs concern all peoples, and they cannot be solved on a national basis. They can only be solved satisfactorily through world-wide, supra-national co-operation.

Co-operation without having to consider boundaries, and opportunities to put capital at the disposal of those who need it for development purposes, would be quite different from what they are at present if universal and comprehensive disarmament could be achieved, and if UN were transformed into a recognized world authority – not only with the task of preserving peace, but also of leading the world towards a high standard of living for all peoples.

In September 1965 an International Symposium on 'Highly Saline and Seawater Irrigation with and without Desalination' was arranged in Rome by the World Academy of Art and Science in co-operation with UNESCO and the Italian Academy of Agriculture and the Research Council. Numbers of lectures were given in which very good results of irrigation with salt water were reported. If these results can be applied on a large scale in desert areas it would be possible to increase food production quite considerably. In view of its global impact on human welfare – as Dr Boyko has expressed it – the World Academy has established a working group on this subject in order to organize close international co-operation for this line of research, and it is to be hoped that the experiments will soon be enlarged on a global scale with the help of the Agencies of UN.

INDEX

GEORGE ALLEN & UNWIN LTD

London: 40 Museum Street, W.C.1

Auckland: P.O. Box 36013, Northcote Central, N.4
Bombay: 15 Graham Road, Ballard Estate, Bombay 1
Barbados: P.O. Box 222, Bridgetown
Buenos Aires: Escritorio 454-459, Florida 165
Calcutta: 17 Chittaranjan Avenue, Calcutta 13
Cape Town: 68 Shortmarket Street
Hong Kong: 105 Wing On Mansion, 26 Hancow Road, Kowloon
Ibadan: P.O. Box 62
Karachi: Karachi Chambers, McLeod Road
Madras: Mohan Mansions, 38c Mount Road, Madras 6
Mexico: Villalongin 32-10, Piso, Mexico 5, D.F.
Nairobi: P.O. Box 4536
New Delhi: 13-14 Asaf Ali Road, New Delhi 1
Ontario: 81 Curlew Drive, Don Mills
Philippines: Manila, P.O. Box 4322
Rio de Janeiro: Caixa Postal 2537-Zc-00
Singapore: 36c Prinsep Street, Singapore 7
Sydney, N.S.W.: Bradbury House, 55 York Street
Tokyo: P.O. Box 26, Kamata

SIR E. JOHN RUSSELL

WORLD POPULATION AND WORLD FOOD SUPPLIES

'A book on world population and food supplies which proves that it is still possible to be both encyclopaedic and readable. . . . Examines the wider and more important historical and social background of each country's agriculture. And a fascinating assembly of knowledge it is.' *The Times*

'It may be doubted whether any other living authority possesses the qualification for the immense survey undertaken by Sir John Russell, abundantly detailed yet remarkably easy to read. He is sparing of judgements, but he supplies the facts, both for advanced and under-developed territories, without which no judgement could be valid . . . will clearly make itself indispensable in this field.' *The Spectator*

'This book, with its wealth of information on the present position and possibility of development of agriculture in the different countries of the world, is a valuable addition to the growing literature on the question of whether the population will grow to a size that the earth will not be able to support.' *Nature*

Royal 8vo. Illustrated.

A HISTORY OF AGRICULTURAL SCIENCE IN GREAT BRITAIN

Sir John Russell completed this book shortly before his death. In it he surveys the history of agricultural science in Great Britain from its tentative beginnings in the seventeenth century until, roughly, the end of the Second World War. It is a subject on which he was uniquely qualified to write. In 1900 he gave up a promising career in Pure Chemistry to become a lecturer at the recently founded Wye Agricultural College. From that time, and more especially from 1912 to 1943 when he was Director of the world-famous Rothamsted Experimental Station, he was in close contact with all the significant developments, many of which he himself helped to initiate.

For the earlier period he was able to use the very fine collection of old agricultural books in the Rothamsted Library. He discusses in some detail the ideas of the great pioneers of agricultural science such as Francis Bacon, Robert Boyle, John Evelyn, Humphrey Davy, Lawes and Gilbert, without neglecting the many lesser writers who made their own contributions to the subject. From 1900 onwards his wide knowledge of the technical literature is supplemented by personal memories of the leading scientists and their work.

This is a book which nobody but Sir John Russell could have written and it is likely to remain the standard work in its field for many years to come.

Demy 8vo.